Collins

2024 GUIDE
to the
NIGHT SKY

Storm Dunlop and Wil Tirion

Published by Collins
An imprint of HarperCollins Publishers
Westerhill Road, Bishopbriggs
Glasgow G64 2QT

HarperCollins Publishers
Macken House,
39/40 Mayor Street Upper,
Dublin 1, D01 C9W8, Ireland
www.harpercollins.co.uk

In association with
Royal Museums Greenwich, the group name for the National Maritime Museum,
Royal Observatory Greenwich, Queen's House and Cutty Sark 2015
www.rmg.co.uk

© HarperCollins Publishers 2023
Text and illustrations © Storm Dunlop and Wil Tirion
Photographs © see acknowledgements page 110

Collins ® is a registered trademark of HarperCollins Publishers Ltd

The contents of this publication are believed correct at the time of printing.
Nevertheless the publisher can accept no responsibility for errors or omissions,
changes in the detail given or for any expense or loss thereby caused.

HarperCollins does not warrant that any website mentioned in this title will be provided uninterrupted,
that any website will be error free, that defects will be corrected, or that the website or the server that
makes it available are free of viruses or bugs. For full terms and conditions please refer to the site
terms provided on the website.

A catalogue record for this book is available from the British Library

ISBN 978-0-00-860430-1

10 9 8 7 6 5 4 3 2 1

Printed in Italy

If you would like to comment on any aspect of this book, please contact us at the above address or online.

Contents

Introduction

The aim of this Guide is to help people find their way around the night sky, by showing how the stars that are visible change from month to month and by including details of various events that occur throughout the year. The objects and events described may be observed with the naked eye, or nothing more complicated than a pair of binoculars.

The conditions for observing naturally vary over the course of the year. During the summer, twilight may persist throughout the night and make it difficult to see the faintest stars. There are three recognized stages of twilight: civil twilight, when the Sun is less than 6° below the horizon; nautical twilight, when the Sun is between 6° and 12° below the horizon; and astronomical twilight, when the Sun is between 12° and 18° below the horizon. Full darkness occurs only when the Sun is more than 18° below the horizon. During nautical twilight, only the very brightest stars are visible. During astronomical twilight, the faintest stars visible to the naked eye may be seen directly overhead, but are lost at lower altitudes. As the diagram shows, during June and most of July full darkness never occurs

at the latitude of London, and at Edinburgh nautical twilight persists throughout the whole night, so at that latitude only the very brightest stars are visible.

Another factor that affects the visibility of objects is the amount of moonlight in the sky. At Full Moon, it may be very difficult to see some of the fainter stars and objects, and even when the Moon is at a smaller phase it may seriously interfere with visibility if it is near the stars or planets in which you are interested. A full lunar calendar is given for each month and may be used to see when nights are likely to be darkest and best for observation.

The celestial sphere

All the objects in the sky (including the Sun, Moon and stars) appear to lie at some indeterminate distance on a large sphere, centred on the Earth. This *celestial sphere* has various reference points and features that are related to those of the Earth. If the Earth's rotational axis is extended, for example, it points to the North and South Celestial Poles, which are thus in line with the North and South Poles on Earth. Similarly, the *celestial*

The duration of twilight throughout the year at London and Edinburgh.

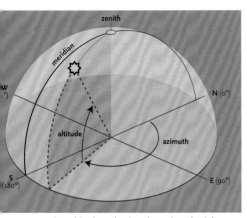

Measuring altitude and azimuth on the celestial sphere.

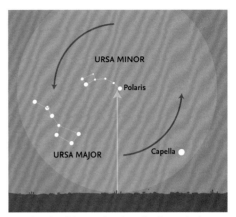

The altitude of the North Celestial Pole equals the observer's latitude.

equator lies in the same plane as the Earth's equator, and divides the sky into northern and southern hemispheres. Because this Guide is written for use in Britain and Ireland, the area of the sky that it describes includes the whole of the northern celestial hemisphere and those portions of the southern that become visible at different times of the year. Stars in the far south, however, remain invisible throughout the year, and are not included.

It is useful to know some of the special terms for various parts of the sky. As seen by an observer, half of the celestial sphere is invisible, below the horizon. The point directly overhead is known as the **zenith**, and the (invisible) one below one's feet as the **nadir**. The line running from the north point on the horizon, up through the zenith and then down to the south point is the **meridian**. This is an important invisible line in the sky, because objects are highest in the sky, and thus easiest to see, when they cross the meridian in the south. Objects are said to **transit**, when they cross this line in the sky.

In this book, reference is frequently made in the text and in the diagrams to the standard compass points around the horizon. The position of any object in the sky may be described by its **altitude** (measured in degrees above the horizon), and its **azimuth** (measured in degrees from north 0°, through east 90°,

south 180° and west 270°). Experienced amateurs and professional astronomers also use another system of specifying locations on the celestial sphere, but that need not concern us here, where the simpler method will suffice.

The celestial sphere appears to rotate about an invisible axis, running between the North and South Celestial Poles. The location (i.e., the altitude) of the Celestial Poles depends entirely on the observer's position on Earth or, more specifically, their latitude. The charts in this book are produced for the latitude of 50°N, so the North Celestial Pole (NCP) is 50° above the northern horizon. The fact that the NCP is fixed relative to the horizon means that all the stars within 50° of the pole are always above the horizon and may, therefore, always be seen at night, regardless of the time of year. This northern circumpolar region is an ideal place to begin learning the sky, and ways to identify the circumpolar stars and constellations will be described shortly.

The ecliptic and the zodiac

Another important line on the celestial sphere is the Sun's apparent path against the background stars – in reality the result of the Earth's orbit around the Sun. This is known as the **ecliptic**. The point where the Sun, apparently moving along the ecliptic, crosses the celestial equator from south to north

The Sun crossing the celestial equator in spring.

is known as the vernal (or spring) equinox, which occurs on March 20. At this time (and at the autumnal equinox, on September 22 or 23, when the Sun crosses the celestial equator from north to south) day and night are almost exactly equal in length. (There is a slight difference, but that need not concern us here.) The vernal equinox is currently located in the constellation of Pisces, and is important in astronomy because it defines the zero point for a system of celestial coordinates, which is, however, not used in this Guide.

The Moon and planets are to be found in a band of sky that extends 8° on either side of the ecliptic. This is because the orbits of the Moon and planets are inclined at various angles to the ecliptic (i.e., to the plane of the Earth's orbit). This band of sky is known as the zodiac and, when originally devised, consisted of twelve **constellations**, all of which were considered to be exactly 30° wide. When the constellation boundaries were formally established by the International Astronomical Union in 1930, the exact extent of most constellations was altered and, nowadays, the ecliptic passes through thirteen constellations.

Because of the boundary changes, the Moon and planets may actually pass through several other constellations that are adjacent to the original twelve.

The constellations

Since ancient times, the celestial sphere has been divided into various constellations, most dating back to antiquity and usually associated with certain myths or legendary people and animals. Nowadays, the boundaries of the constellations have been fixed by international agreement and their names (in Latin) are largely derived from Greek or Roman originals. Some of the names of the most prominent stars are of Greek or Roman origin, but many are derived from Arabic names. Many bright stars have no individual names and, for many years, stars were identified by terms such as 'the star in Hercules' right foot'. A more sensible scheme was introduced by the German astronomer Johannes Bayer in the early seventeenth century. Following his scheme – which is still used today – most of the brightest stars are identified by a Greek letter followed by the genitive form of the constellation's Latin name. An example is the Pole Star, also known as Polaris and α Ursae Minoris (abbreviated α UMi). The Greek alphabet is shown on page 109 with a list of all the constellations that may be seen from latitude 50°N, together with abbreviations, their genitive forms and English names. Other naming schemes exist for fainter stars, but are not used in this book.

Asterisms

Apart from the constellations (88 of which cover the whole sky), certain groups of stars, which may form a part of a larger constellation or cross several constellations, are readily recognizable and have been given individual names. These groups are known as *asterisms*, and the most famous (and well-known) is the 'Plough', the common name for the seven brightest stars in the constellation of Ursa Major, the Great Bear. The names and details of some asterisms mentioned in this book are given in the list on page 110.

Magnitudes

The brightness of a star, planet or other body is frequently given in magnitudes (mag.). This is a mathematically defined scale where larger numbers indicate a fainter object. The scale extends beyond the zero point to negative numbers for very bright objects. (Sirius, the brightest star in the sky is mag. -1.4.) Most observers are able to see stars to about mag. 6, under very clear skies.

The Moon

Although the daily rotation of the Earth carries the sky from east to west, the Moon gradually moves eastwards by approximately its diameter (about half a degree) in an hour. Normally, in its orbit around the Earth, the Moon passes above or below the direct line between Earth and Sun (at New Moon) or outside the area obscured by the Earth's shadow (at Full Moon). Occasionally, however, the three bodies are more-or-less perfectly aligned to give an *eclipse*: a solar eclipse at New Moon or a lunar eclipse at Full Moon. Depending on the exact circumstances, a solar eclipse may be merely partial (when the Moon does not cover the whole of the Sun's disk); annular (when the Moon is too far from Earth in its orbit to appear large enough to hide the whole of the Sun); or total. Total and annular eclipses are visible from very restricted areas of the Earth, but partial eclipses are normally visible over a wider area.

Somewhat similarly, at a lunar eclipse, the Moon may pass through the outer zone of the Earth's shadow, the **penumbra** (in a penumbral eclipse, which is not generally perceptible to the naked eye), so that just part of the Moon is within the darkest part of the Earth's shadow, the **umbra** (in a partial eclipse); or completely within the umbra (in a total eclipse). Unlike solar eclipses, lunar eclipses are visible from large areas of the Earth.

Occasionally, as it moves across the sky, the Moon passes between the Earth and individual planets or distant stars, giving rise to an **occultation**. As with solar eclipses, such occultations are visible from restricted areas of the world.

The planets

Because the planets are always moving against the background stars, they are treated in some detail in the monthly pages and information is given when they are close to other planets, the Moon or any of five bright stars that lie near the ecliptic. Such events are known as **appulses** or, more frequently, as **conjunctions**. (There are technical differences in the way these terms are defined – and should be used – in astronomy, but these need not concern us here.) The positions of the planets are shown for every month on a special chart of the ecliptic.

The term conjunction is also used when a planet is either directly behind or in front of the Sun, as seen from Earth. (Under normal circumstances it will then be invisible.) The conditions of most favourable visibility depend on whether the planet is one of the two known as **inferior planets** (Mercury and Venus) or one of the three **superior planets** (Mars, Jupiter

Inferior planet.

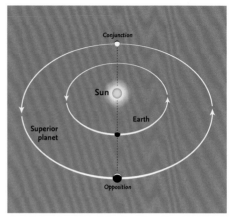

Superior planet.

and Saturn) that are covered in detail. (Some details of the fainter superior planets, Uranus and Neptune, are included in this Guide, and special charts are given on page 25.)

The inferior planets are most readily seen at eastern or western **elongation**, when their angular distance from the Sun is greatest. For superior planets, they are best seen at **opposition**, when they are directly opposite the Sun in the sky, and cross the meridian at local midnight.

It is often useful to be able to estimate angles on the sky, and approximate values may be obtained by holding one hand at arm's length. The various angles are shown in the diagram, together with the separations of the various stars in the Plough.

Meteors

At some time or other, nearly everyone has seen a **meteor** – a 'shooting star' – as it flashed across the sky. The particles that cause meteors – known technically as 'meteoroids' – range in size from that of a grain of sand (or even smaller) to the size of a pea. On any night of the year there are occasional meteors, known as **sporadics**, that may travel in any direction. These occur at a rate that is normally between three and eight in an hour. Far more important, however, are **meteor showers**, which occur at fixed periods of the year, when the

Earth encounters a trail of particles left behind by a comet or, very occasionally, by a minor planet (asteroid). Meteors always appear to diverge from a single point on the sky, known as the **radiant**, and the radiants of major showers are shown on the charts. Meteors that come from a circular area 8° in diameter around the radiant are classed as belonging to the particular shower. All others that do not come from that area are sporadics (or, occasionally from another shower that is active at the same time). A list of the major meteor showers is given on page 31.

Although the positions of the various shower radiants are shown on the charts, looking directly at the radiant is not the most effective way of seeing meteors. They are most likely to be noticed if one is looking about 40–45° away from the radiant position. (This is approximately two hand-spans as shown in the diagram for measuring angles.)

Other objects

Certain other objects may be seen with the naked eye under good conditions. Some were given names in antiquity – Praesepe is one example – but many are known by what are called 'Messier numbers', the numbers in a catalogue of nebulous objects compiled by Charles Messier in the late eighteenth century. Some, such as the Andromeda Galaxy, M31, and the Orion Nebula, M42, may be seen

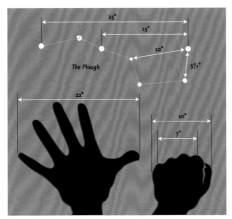

Measuring angles in the sky.

Meteor shower (showing the April Lyrid radiant).

by the naked eye, but all those given in the list will benefit from the use of binoculars. Apart from galaxies, such as M31, which contain thousands of millions of stars, there are also two types of cluster: open clusters, such as M45, the Pleiades, which may consist of a few dozen to some hundreds of stars; and globular clusters, such as M13 in Hercules, which are spherical concentrations of many thousands of stars. One or two gaseous nebulae, consisting of gas illuminated by stars within them, are also visible. The Orion Nebula, M42, is one, and is illuminated by the group of four stars, known as the Trapezium, which may be seen within it by using a good pair of binoculars.

Some interesting objects.

Messier / NGC	Name	Type	Constellation	Maps (months)
—	Hyades	open cluster	Taurus	Sep. – Mar.
—	Double Cluster	open cluster	Perseus	All year
—	Melotte 111 (Coma Cluster)	open cluster	Coma Berenices	Jan. – Aug.
M3	—	globular cluster	Canes Venatici	Jan. – Sep.
M4	—	globular cluster	Scorpius	May – Aug.
M8	Lagoon Nebula	gaseous nebula	Sagittarius	Jun. – Aug.
M11	Wild Duck Cluster	open cluster	Scutum	May – Oct.
M13	Hercules Cluster	globular cluster	Hercules	Feb. – Nov.
M15	—	globular cluster	Pegasus	Jun. – Dec.
M20	Trifid Nebula	gaseous nebula	Sagittarius	Jun. – Aug.
M22	—	globular cluster	Sagittarius	Jun. – Sep.
M27	Dumbbell Nebula	planetary nebula	Vulpecula	May – Dec.
M31	Andromeda Galaxy	galaxy	Andromeda	All year
M35	—	open cluster	Gemini	Oct. – May
M42	Orion Nebula	gaseous nebula	Orion	Nov. – Mar.
M44	Praesepe	open cluster	Cancer	Nov. – Jun.
M45	Pleiades	open cluster	Taurus	Aug. – Apr.
M57	Ring Nebula	planetary nebula	Lyra	Apr. – Dec.
M67	—	open cluster	Cancer	Dec. – May
NGC 752	—	open cluster	Andromeda	Jul. – Mar.
NGC 3242	Ghost of Jupiter	planetary nebula	Hydra	Feb. – May

The Northern Circumpolar Constellations

The northern circumpolar stars are the key to starting to identify the constellations. For anyone in the northern hemisphere they are visible at any time of the year, and nearly everyone is familiar with the seven stars of the Plough – known as the Big Dipper in North America – an asterism that forms part of the large constellation of **Ursa Major** (the Great Bear).

Ursa Major

Because of the movement of the stars caused by the passage of the seasons, Ursa Major lies in different parts of the evening sky at different periods of the year. The diagram below shows its position at the beginning of the four main seasons. The seven stars of the Plough remain visible throughout the year anywhere north of latitude 40°N. Even at the latitude (50°N) for which the charts in this book are drawn, many of the stars in the southern portion of the constellation of Ursa Major are hidden below the horizon for part of the year or (particularly in late summer) cannot be seen late in the night.

Polaris and Ursa Minor

The two stars **Dubhe** and **Merak** (α and β Ursae

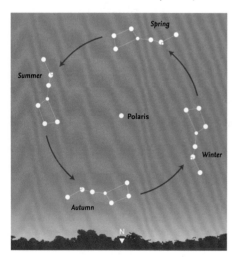

Majoris, respectively), farthest from the 'tail' are known as the 'Pointers'. A line from Merak to Dubhe, extended about five times their separation, leads to the Pole Star, **Polaris**, or α Ursae Minoris. All the stars in the northern sky appear to rotate around it. There are five main stars in the constellation of **Ursa Minor**, and the two farthest from the Pole, **Kochab** and **Pherkad** (β and γ Ursae Minoris, respectively), are known as 'The Guards'.

Cassiopeia

On the opposite of the North Pole from Ursa Major lies **Cassiopeia**. It is highly distinctive, appearing as five stars forming a letter 'W' or 'M' depending on its orientation. Provided the sky is reasonably clear of clouds, you will nearly always be able to see either Ursa Major or Cassiopeia, and thus be able to orientate yourself on the sky.

To find Cassiopeia, start with **Alioth** (ε Ursae Majoris), the first star in the tail of the Great Bear. A line from this star extended through Polaris points directly towards γ Cassiopeiae, the central star of the five.

Cepheus

Although the constellation of **Cepheus** is fully circumpolar, it is not nearly as well-known as Ursa Major, Ursa Minor or Cassiopeia, partly because its stars are fainter. Its shape is rather like the gable-end of a house. The line from the Pointers through Polaris, if extended, leads to **Errai** (γ Cephei) at the 'top' of the 'gable'. The brightest star, **Alderamin** (α Cephei) lies in the Milky Way region, at the 'bottom right-hand corner' of the figure.

Draco

The constellation of **Draco** consists of a quadrilateral of stars, known as the 'Head of Draco' (and also the 'Lozenge'), and a long chain of stars forming the neck and body of the dragon. To find the Head of Draco, locate the two stars **Phecda** and **Megrez** (γ and δ Ursae Majoris) in the Plough, opposite the Pointers.

The stars and constellations inside the circle are always above the horizon, seen from our latitude.

Extend a line from Phecda through Megrez by about eight times their separation, right across the sky below the Guards in Ursa Minor, ending at **Grumium** (ξ Draconis) at one corner of the quadrilateral. The brightest star, **Eltanin** (γ Draconis) lies farther to the south. From the head of Draco, the constellation first runs northwest to **Altais** (δ Draconis) and ε Draconis, then doubles back southwards before winding its way through **Thuban** (α Draconis) before ending at **Giausar** (λ Draconis) between the Pointers and Polaris.

The Winter Constellations

The winter sky is dominated by several bright stars and distinctive constellations. The most conspicuous constellation is **Orion**, the main body of which has an hour-glass shape. It straddles the celestial equator and is thus visible from anywhere in the world. The three stars that form the 'Belt' of Orion point down towards the southeast and to **Sirius** (α Canis Majoris), the brightest star in the sky. **Mintaka** (δ Orionis), the star at the northeastern end of the Belt, farthest from Sirius, actually lies just slightly south of the celestial equator.

A line from **Bellatrix** (γ Orionis) at the 'top right-hand corner' of Orion, through **Aldebaran** (α Tauri), past the 'V' of the Hyades cluster, points to the distinctive cluster of bright blue stars known as the **Pleiades**, or the 'Seven Sisters'. Aldebaran is one of the five bright stars that may sometimes be occulted (hidden) by the Moon. Another line from Bellatrix, through

Betelgeuse (α Orionis), if carried right across the sky, points to the constellation of **Leo**, a prominent constellation in the spring sky.

Six bright stars in six different constellations: **Capella** (α Aurigae), **Aldebaran** (α Tauri), **Rigel** (β Orionis), **Sirius** (α Canis Majoris), **Procyon** (α Canis Minoris) and **Pollux** (β Gemini) form what is sometimes known as the 'Winter Hexagon'. Pollux is accompanied to the northwest by the slightly fainter star of **Castor** (α Gemini), the second 'Twin'.

In a counterpart to the famous 'Summer Triangle', an almost perfect equilateral triangle, the 'Winter Triangle', is formed by Betelgeuse (α Orionis), Sirius (α Canis Majoris) and Procyon (α Canis Minoris).

Several of the stars in this region of the sky show distinctive tints: Betelgeuse (α Orionis) is reddish, Aldebaran (α Tauri) is orange and Rigel (β Orionis) is blue-white.

The Spring Constellations

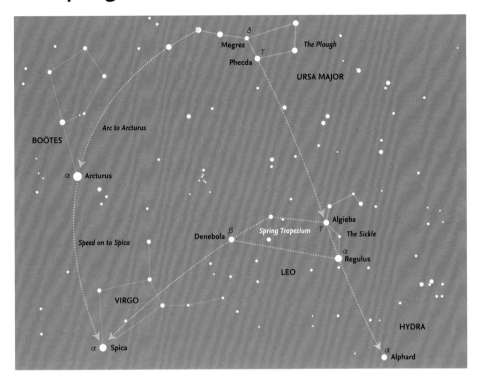

The most prominent constellation in the spring sky is the zodiacal constellation of **Leo**, and its brightest star, **Regulus** (α Leonis), which may be found by extending a line from Megrez and Phecda (δ and γ Ursae Majoris, respectively) – the two stars on the opposite side of the bowl of the Plough from the Pointers – down to the southeast. Regulus forms the 'dot' of the 'backward question mark' known as 'the Sickle'. Regulus, like Aldebaran in Taurus is one of the bright stars that lie close to the ecliptic, and that are occasionally occulted by the Moon. The same line from Ursa Major to Regulus, if continued, leads to **Alphard** (α Hydrae), the brightest star in **Hydra**, the largest of the 88 constellations.

The shape formed by the body of Leo is sometimes known as the 'Spring Trapezium'. At the other end of the constellation from Regulus is **Denebola** (β Leonis), and the line forming the back of the constellation through Denebola points to the bright star **Spica** (α Virginis) in the constellation of **Virgo**. A saying that helps to locate Spica is well-known to astronomers: 'Arc to Arcturus and then speed on to Spica.' This suggests following the arc of the tail of Ursa Major to Arcturus and then on to Spica. **Arcturus** (α Boötis) is actually the brightest star in the northern hemisphere of the sky. (Although other stars, such as Sirius, are brighter, they are all in the southern hemisphere.) Overall, the constellation of **Boötes** is sometimes described as 'kite-shaped' or 'shaped like the letter P'.

Although Spica is the brightest star in the zodiacal constellation of Virgo, the rest of the constellation is not particularly distinct, consisting of a rough quadrilateral of moderately bright stars and some fainter lines of stars extending outwards.

The Summer Constellations

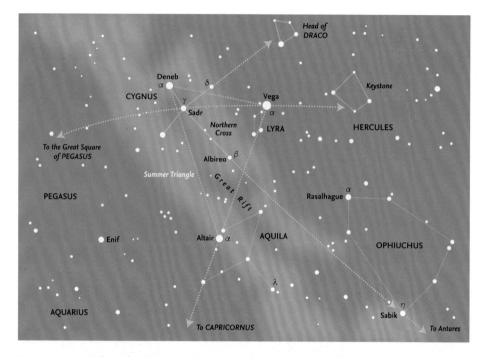

On summer nights, the three bright stars **Deneb** (α Cygni), **Vega** (α Lyrae) and **Altair** (α Aquilae) form the striking 'Summer Triangle'. The constellations of **Cygnus** (the Swan) and **Aquila** (the Eagle) represent birds 'flying' down the length of the Milky Way. This part of the Milky Way contains the **Great Rift**, an elongated dark region, where the light from distant stars is obscured by intervening dust. The dark Rift is clearly visible even to the naked eye.

The most prominent stars of Cygnus are sometimes known as the 'Northern Cross' (as a counterpart to the 'Southern Cross' – the constellation of Crux – in the southern hemisphere). The central line of Cygnus through **Albireo** (β Cygni), extended well to the southwest, points to **Sabik** (η Ophiuchi) in the large, sprawling constellation of **Ophiuchus** (the Serpent Bearer) and beyond to **Antares** (α Scorpii) in the constellation of **Scorpius**. Like Cepheus, the shape of Ophiuchus somewhat resembles the gable-end of a house, and the

brightest star **Rasalhague** (α Ophiuchi) is at the 'apex' of the 'gable'.

A line from the central star of Cygnus, **Sadr** (γ Cygni) through δ Cygni, in the northwestern 'wing' points towards the Head of Draco, and is another way of locating that part of the constellation. Another line from Sadr to Vega indicates the central portion, 'The Keystone', of the constellation of **Hercules**. An arc through the same stars, in the opposite direction, points towards the constellation of **Pegasus**, and more specifically to the 'Great Square of Pegasus'.

Aquila is less conspicuous than Cygnus and consists of a diamond shape of stars, representing the body and wings of the eagle, together with a rather faint star, λ Aquilae, marking the 'head'. **Lyra** (the Lyre) mainly consists of Vega (α Lyrae) and a small quadrilateral of stars to its southeast. Continuation of a line from Vega through Altair indicates the zodiacal constellation of **Capricornus**.

The Autumn Constellations

During the autumn season, the most striking feature is the 'Great Square of Pegasus', an almost perfect rectangle on the sky, forming the main body of the constellation of **Pegasus**. However, the star at the northeastern corner, **Alpheratz**, is actually α Andromedae, and part of the adjacent constellation of **Andromeda**. A line from **Scheat** (β Pegasi) at the northeastern corner of the Square, through **Matar** (η Pegasi), points in the general direction of Cygnus. A crooked line of stars leads from **Markab** (α Pegasi) through **Homam** (ζ Pegasi) and **Biham** (θ Pegasi) to **Enif** (ε Pegasi). A line from Markab through the last star in the Square, **Algenib** (γ Pegasi) points in the general direction of the five stars, including **Menkar** (α Ceti) that form the 'tail' of the constellation of **Cetus** (the Whale). A ring of seven stars lying below the southern side of the Great Square is known as 'the Circlet', part of the constellation of **Pisces** (the Fishes).

Extending the line of the western side of the Great Square towards the south leads to the isolated bright star, **Fomalhaut** (α Piscis Austrini), in the Southern Fish. Following the line of the eastern side of the Great Square towards the north leads to Cassiopeia while, in the other direction, it points towards **Diphda** (β Ceti), which is actually the brightest star in Cetus.

Three bright stars leading northeast from Alpheratz form the main body of the constellation of **Andromeda**. Continuation of that line leads towards the constellation of **Perseus** and **Mirfak** (α Persei). Running southwards from Mirfak is a chain of stars, one of which is the famous variable star **Algol** (β Persei). Farther east, an arc of stars leads to the prominent cluster of the **Pleiades**, in the constellation of **Taurus**.

Between Andromeda and Cetus lie the two small constellations of **Triangulum** (the Triangle) and **Aries** (the Ram).

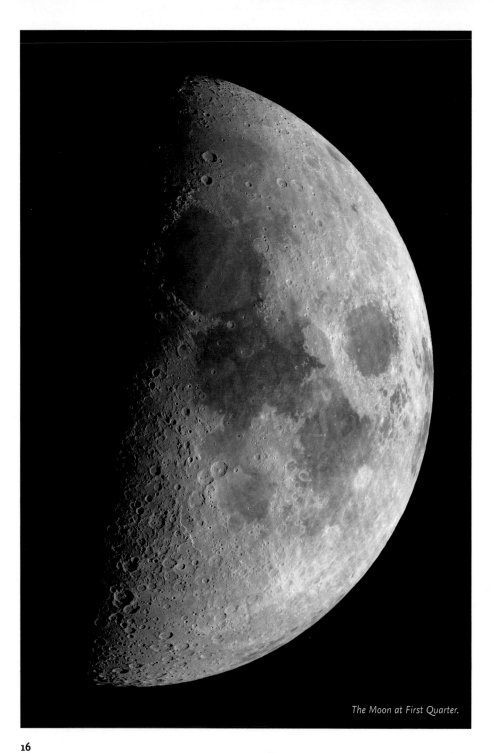

The Moon at First Quarter.

The Moon

The monthly pages include diagrams showing the phase of the Moon for every day of the month, and also indicate the day in the **lunation** (or **age** of the Moon), which begins at New Moon. Although the main features of the surface – the light highlands and the dark maria (seas) – may be seen with the naked eye, far more features may be detected with the use of binoculars or any telescope. The many craters are best seen when they are close to the **terminator** (the boundary between the illuminated and the non-illuminated areas of the surface), when the Sun rises or sets over any particular region of the Moon and the crater walls or central peaks cast strong shadows. Most features become difficult to see at Full Moon, although this is the best time to see the bright ray systems surrounding certain craters. Accompanying the Moon map on the following pages is a list of prominent features, including the days in the lunation when features are normally close to the terminator and thus easiest to see. A few bright features

such as Linné and Proclus, visible when well illuminated, are also listed. One feature, Rupes Recta (the Straight Wall) is readily visible only when it casts a shadow with light from the east, appearing as a light line when illuminated from the opposite direction.

The dates of visibility vary slightly through the effects of **libration**. Because the Moon's orbit is inclined to the Earth's equator and also because it moves in an ellipse, the Moon appears to rock slightly from side to side (and nod up and down). Features near the **limb** (the edge of the Moon) may vary considerably in their location and visibility. (This is easily noticeable with Mare Crisium and the craters Tycho and Plato.) Another effect is that at crescent phases before and after New Moon, the normally non-illuminated portion of the Moon receives a certain amount of light, reflected from the Earth. This **Earthshine** may enable certain bright features (such as Aristarchus, Kepler and Copernicus) to be detected even though they are not illuminated by sunlight.

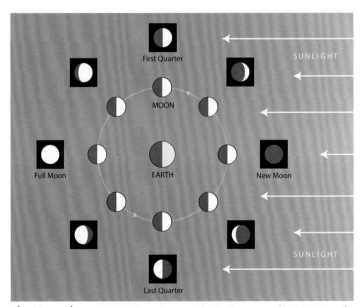

The Moon phases. *During its orbit around the Earth we see different portions of the illuminated side of the Moon's surface.*

Map of the Moon

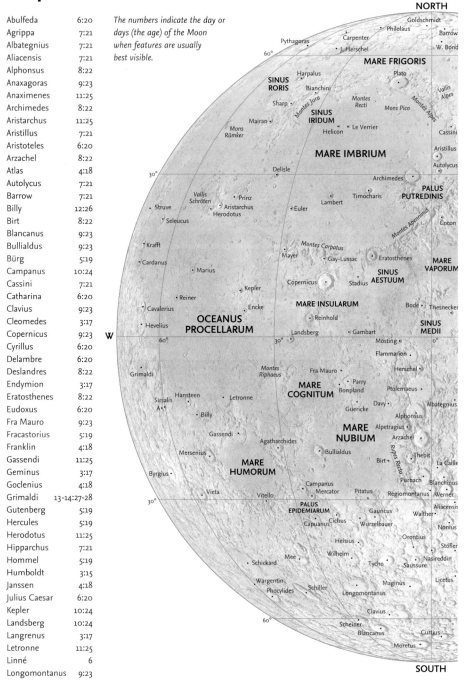

The numbers indicate the day or days (the age) of the Moon when features are usually best visible.

Goldschmidt · Meton
Barrow
W. Bond · Arnold
60°

MARE
FRIGORIS
Endymion
Vallis Alpes · Aristoteles · Mercurius
Montes Alpes · Eudoxus · Bürg · Hercules · Atlas
LACUS MORTIS
Cassini · Cepheus
LACUS SOMNIORUM · Franklin
Montes Caucasus · Gauss
Aristillus · Geminus · Berosus
Archimedes · Autolycus · Posidonius · Hahn
MARE SERENITATIS · Chacornac · Burckhardt
30°
PALUS PUTREDINIS · Linné · Cleomedes
Montes Hoemus · Le Monnier
Cocon · Bessel
Montes Apenninus · Macrobius
Manilius · Menelaus · Plinius · Vetruvius
MARE CRISIUM
MARE VAPORUM
PALUS SOMNI · Picard
Hyginus · MARE TRANQUILLITATIS · Condorcet
Bode · Triesnecker · Agrippa · Arago · Firmicus · MARE UNDARUM
SINUS MEDII · Godin · Ritter · Taruntius · Apollonius
Rhaeticus · Sabine · MARE SPUMANS
0° · 30° · 60° · E
Flammarion · Delambre · A · Messier
Herschel · Hypatia · Torricelli · MARE FECUNDITATIS
Hipparchus · Capella · Gutenberg · Langrenus
Ptolemaeus · Isidorus
Albategnius · Theophilus · Goclenius · Kapteyn
Alphonsus · Abulfeda · Cyrillus · Mädler
Argelander · MARE NECTARIS · Colombo · Vendelinus
Arzachel · Airy · Almanon · Tacitus · Catharina · Beaumont
Geber · Fracastorius
Abenezra · Santbech
Thebit · La Caille · Playfair · Azophi
Purbach · Blanchinus · Sacrobosco · Petavius
Apianus · Humboldt
Regiomontanus · Werner · Pontanus · Piccolomini · Snellius
Walther · Aliacensis · Zagut · Reichenbach · 30°
Nonius · Rabbi Levi · Stevinus
Orontius · Büsching · Furnerius
Stöfler · Buch · Riccius · Metius · Rheita
Nasireddin · Maurolycus · Fabricius · Vallis Rheita
Saussure
Maginus · Licetus · Barocius
Cuvier · Baco · Pitiscus
Curtius · Hommel · Vlacq
Jacobi
Curtius · Manzinus · 60°
Moretus · MARE AUSTRALE

Eclipses in 2024

Lunar eclipses

There are two lunar eclipses in 2024, one penumbral on March 25, with one partial eclipse on September 18. The penumbral eclipse is of minimal interest, mainly because to the naked eye it will show little variation in the Moon. During the first eclipse, on March 25, a very small portion of the Moon may just graze the umbra. Maximum eclipse is at 07:12. The event (such as it is) will be visible in its totality from North America. During the second (partial) eclipse, on September 18, a slightly larger portion of the Moon will enter the umbra, beginning at 02:14 and leaving at 03:16, with maximum eclipse at 02:44. The middle and end of this eclipse will be just visible from the eastern seaboard of the United States.

Solar eclipses

There are two solar eclipses in 2024. The total eclipse of April 8 is the most significant. The path of totality passes across the southern United States and is shown on the map below. Maximum eclipse actually occurs in Mexico at 18:17, where the duration is 4 minutes 28.1 seconds, but the path of totality runs all the way from Texas to Nova Scotia and Newfoundland in Canada, ending in the North Atlantic.

The second solar eclipse on October 2 is an annular eclipse, mainly visible over the Pacific Ocean, although the end of the track does pass across southern Chile and Argentina. Although most of the track passes over the open ocean, it will cross Rapa Nui (Easter Island, at 27°7′S, 109°22′W) where the annular phase will be visible and last over six minutes. Maximum eclipse occurs at 18:46.

The total solar eclipse of April 8, 2024. The path of totality sees maximum eclipse duration occur in Mexico, before the track enters the United States in Texas. The path runs right across the United States, with the ground track ending in Nova Scotia and Newfoundland in Canada.

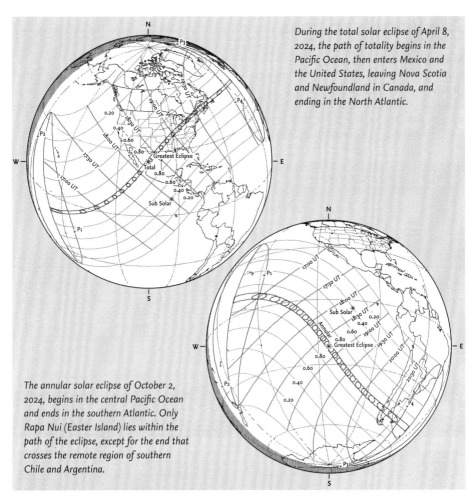

During the total solar eclipse of April 8, 2024, the path of totality begins in the Pacific Ocean, then enters Mexico and the United States, leaving Nova Scotia and Newfoundland in Canada, and ending in the North Atlantic.

The annular solar eclipse of October 2, 2024, begins in the central Pacific Ocean and ends in the southern Atlantic. Only Rapa Nui (Easter Island) lies within the path of the eclipse, except for the end that crosses the remote region of southern Chile and Argentina.

An annular eclipse of the Sun, photographed from California, towards the end of the eclipse at sunset over the Pacific.

The Planets in 2024

Mercury and Venus

Although **Mercury** comes to greatest elongation on seven occasions in 2024, on only four is it reasonably visible above the horizon: January 12, March 24, September 5 and December 25. On one of these (March 24) Jupiter is nearby and on January 12, Venus and Mars are in the same area. **Venus** comes to greatest elongation on October 23, 2023 and January 10, 2025, but not in 2024.

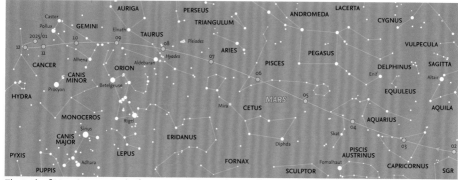

The path of Mars in 2024. In the first months it is too close to the Sun to be readily observed.

Mars

Because its orbit lies outside that of the Earth, so taking longer to complete an orbit, Mars does not come to opposition every year. Instead it often remains in the sky for many months at a time, moving slowly along the ecliptic. This is the case in 2024, when there is no opposition, but the planet is visible from March to December, as shown in the large chart here. (In January and February it is too close to the Sun to be readily visible.) It is mag. 1.3 in **Capricornus** in March, but slowly increases to mag. 1.0 in June and July, in **Pisces** and **Aries**. It ends the year in **Cancer** at mag. -1.2.

Oppositions occur during a period of retrograde motion, when the planet appears to move westwards against the pattern of distant stars. There is no opposition in 2024. Mars actually begins to retrograde on December 9, 2024, and comes to its next opposition on January 16, 2025.

Because of its eccentric orbit, which carries it at very differing distances from the Sun (and Earth), not all oppositions of Mars are equally favorable for observation. The relative positions of Mars and the Earth are shown here. It will be seen that the opposition of 2018 was very close and thus favorable for observation, and that of 2020 was also reasonably good. By comparison, opposition in 2027 will be at a far greater distance, so the planet will appear much smaller.

This image of Mars was obtained by Damian Peach with a 14-inch Celestron telescope on September 30, 2020 when Mars was magnitude -2.5. South is at the top.

The oppositions of Mars between 2018 and 2033. As the illustration shows, there is no opposition in 2024.

The path of Jupiter in 2024. Jupiter comes to opposition on December 7. Background stars are shown down to magnitude 6.5.

Jupiter and Saturn

In 2024, **Jupiter** starts the year in **Aries**, begining in the west. It moves into **Taurus** in late April. It begins to retrograde on October 9 and comes to opposition, still in Taurus, on December 7 at mag. -2.8. It continues retrograde motion to the end of the year. **Saturn** is in **Aquarius** and stays there for the whole year. It starts retrograde motion on 4 July and comes to opposition at mag. 0.6 on September 8. It continues slow retrograde motion until November 16. At the end of the year it is mag. 1.1.

Jupiter's four large satellites are readily visible in binoculars. Not all four are visible all the time, sometimes hidden behind the planet or invisible in front of it. **Io**, the closest to Jupiter, orbits in just under 1.8 days, and **Callisto**, the farthest away, takes about 16.7 days. In between are **Europa** (*c.* 3.6 days) and **Ganymede**, the largest (*c.* 7.1 days). The diagram (below) shows the satellites' motions around the time of opposition.

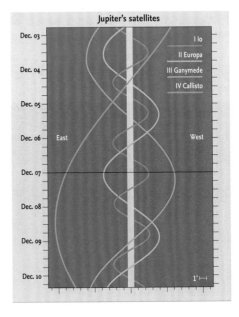

The path of Saturn in 2024. Saturn comes to opposition on September 8. Background stars are shown down to magnitude 6.5.

Uranus and Neptune

Uranus begins the year in **Aries** at mag. 5.7, and moves into **Taurus** in late May. It begins retrograde motion in early September and is at opposition (mag. 5.6) on November 17. It continues retrograde motion and is on the **Taurus/Aries** border at the end of the year.

Neptune begins the year just north of the border between **Aquarius** and **Pisces**. It starts its retrograde motion in early July, and is at opposition (mag. 7.8) on September 21. Still in **Pisces** (at mag. 7.9), it reverts to direct motion at the very end of the year.

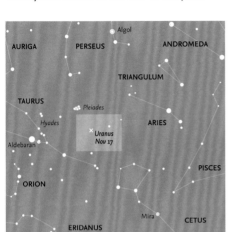

Uranus is in Taurus at the time of its opposition, on November 17. The boxed area is shown in more detail to the right.

The path of Uranus in 2024. Uranus comes to opposition on November 17. All stars brighter than magnitude 7.5 are shown.

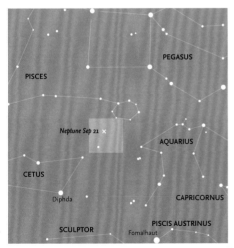

In 2024, Neptune is to be found in the southern part of Pisces. The planet comes to opposition on September 21.

The path of Neptune in 2024. It is in the constellation of Pisces for the whole year. All stars down to magnitude 8.5 are shown.

Minor Planets in 2024

Several minor planets rise above magnitude 9 in 2024, and four come to opposition at, or above mag. 9.0. These are **(3) Juno** in **Leo** on March 3 when it is mag. 8.7; **(1) Ceres** in **Sagittarius** on July 6, at mag. 7.3; **(7) Iris** in **Aquarius** on August 6 (mag. 8.1) and **(15) Eunomia** in **Auriga** at mag. 8.0 on December 14.

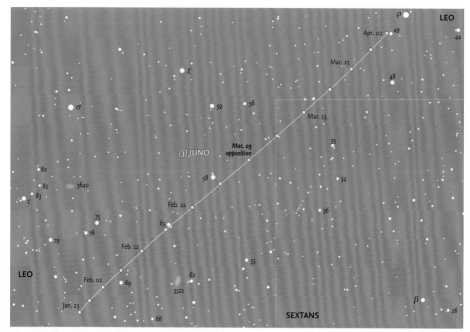

The path of the minor planet (3) Juno around its opposition on March 3 (mag. 8.7). Background stars are shown down to magnitude 9.5.

The path of the minor planet (1) Ceres around its opposition on July 6 (mag. 7.3). Background stars are shown down to magnitude 8.5.

The path of the minor planet (7) Iris around its opposition on August 6 (mag. 8.1). Background stars are shown down to magnitude 9.0.

The path of the minor planet (15) Eunomia around its opposition on December 14 (mag. 8.0). Background stars are shown down to magnitude 9.0.

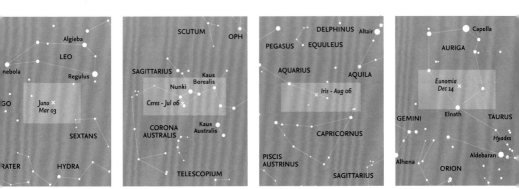

The four small maps at the bottom of this page show the areas covered by the larger maps in a lighter blue. A white cross marks the position of the minor planet at the day of its opposition.

Comets in 2024

Although comets may occasionally become very striking objects in the sky, as was the case with Comet C/2020 F3 NEOWISE in 2020, their occurrence and particularly the existence or length of any tail and their overall magnitude are notoriously difficult to predict. Naturally, it is only possible to predict the return of periodic comets (whose names have the prefix 'P'). Many comets appear unexpectedly (these have names with the prefix 'C'). Bright, readily visible comets such as C/1995 Y1 Hyakutake, C/1995 O1 Hale-Bopp, C/2006 P1 McNaught or C/2020 F3 NEOWISE are rare. Most periodic comets are faint and only a very small number

ever become bright enough to be easily visible with the naked eye or with binoculars.

Comet 62P/Tsuchinshan 1, visible in November and December 2023, may continue to be seen (although fading) in early 2024.

The comet most likely to become readily visible in 2024 is Comet 12P/Pons-Brooks, which may reach mag. 5 during late March or even mag. 4 at the beginning of April. This is a 'Halley-type' comet, with an orbital period of 68.82 years. It last came to perihelion on May 22, 1954.

Another comet, C/2021 S3 (PanSTARRS) may rise as high as mag. 7 during April, but is unlikely to become a prominent object.

The path of Comet Pons-Brooks from January 22 to April 16, 2024. It rises above magnitude 4 in April. Background stars are shown down to magnitude 6.0.

The path of Comet Pons-Brooks from April 11 to June 30, 2024. It rises above magnitude 4 in April. Background stars are shown down to magnitude 6.0.

The Comet C/2020 F3 NEOWISE photographed by Nick James on July 17, 2020, showing the light-coloured dust tail, together with the blue ion tail.

Introduction to the Month-by-Month Guide

The monthly charts

The pages devoted to each month contain a pair of charts showing the appearance of the night sky, looking north and looking south. The charts (as with all the charts in this book) are drawn for the latitude of 50°N, so observers farther north will see slightly more of the sky on the northern horizon, and slightly less on the southern. These areas are, of course, those most likely to be affected by poor observing conditions caused by haze, mist or smoke. In addition, stars close to the horizon are always dimmed by atmospheric absorption, so sometimes the faintest stars marked on the charts may not be visible.

The three times shown for each chart require a little explanation. The charts are drawn to show the appearance at 23:00 GMT for the 1st of each month. The same appearance will apply an hour earlier (22:00 GMT) on the 15th, and yet another hour earlier (21:00 GMT) at the end of the month (shown as the 1st of the following month). GMT is identical to the Universal Time (UT) used by astronomers around the world. In Europe, Summer Time is introduced in March, so the March charts apply to 23:00 GMT on March 1, 22:00 GMT on March 15, but 22:00 BST (British Summer Time) on April 1. The change back from Summer Time (in Europe) occurs in October, so the charts for that month apply to 00:00 BST for October 1, 23:00 BST for October 15, and 21:00 GMT for November 1.

The charts may be used for earlier or later times during the night. To observe two hours earlier, use the charts for the preceding month; for two hours later, the charts for the next month.

Meteors

Details of specific meteor showers are given in the months when they come to maximum, regardless of whether they begin or end in other months. Note that not all the respective radiants are marked on the charts for that

particular month, because the radiants may be below the horizon, or lie in constellations that are not readily visible during the month of maximum. For this reason, special charts for the Eta (η) and Delta (δ) Aquariids (May and July, respectively) and the Leonids (November) are given here. As explained earlier, however, meteors from such showers may still be seen, because the most effective region for seeing meteors is some 40–45° away from the radiant, and that area of sky may well be above the horizon. A table of the best meteor showers visible during the year is also given here. The rates given are based on the properties of the meteor streams, and are those that an experienced observer might see under ideal conditions. Generally, the observed rates will be far less.

Shower	Dates of activity 2024	Date of maximum 2024	Possible hourly rate
Quadrantids	December 28 to January 12	January 3–4	110
April Lyrids	April 14–30	April 22–23	18
η-Aquariids	April 19 to May 28	May 6	50
α-Capricornids	July 3 to August 15	July 30	5
Southern δ-Aquariids	July 12 to August 23	July 30	25
Perseids	July 17 to August 24	August 12–13	100
α-Aurigids	August 28 to September 5	September 1	6
Southern Taurids	September 10 to November 20	October 10–11	5
Orionids	October 2 to November 7	October 21–22	25
Draconids	October 6–10	October 8–9	10
Northern Taurids	October 20 to December 10	November 12–13	5
Leonids	November 6–30	November 18	10
Geminids	December 4–20	December 14–15	150
Ursids	December 17–26	December 23	10

Meteors that are brighter than magnitude -4 (approximately the maximum magnitude reached by Venus) are known as **fireballs** or **bolides**. Examples are shown on pages 32, 35 and 77. Fireballs sometimes cause sonic booms that may be heard some time after the meteor is seen.

The photographs

As an aid to identification – especially as some people find it difficult to relate charts to the actual stars they see in the sky – one or more photographs of constellations visible in certain specific months are included. It should be noted, however, that because of the limitations of the photographic and printing processes, and the differences between the sensitivity of different individuals to faint starlight (especially in their ability to detect different colours), and the degree to which they have become adapted to the dark, the apparent brightness of stars in the photographs will not necessarily precisely match that seen by any one observer.

The Moon calendar

The Moon calendar is largely self-explanatory. It shows the phase of the Moon for every day of the month, with the exact times (in Universal Time) of New Moon, First Quarter, Full Moon and Last Quarter. Because the times are calculated from the Moon's actual orbital parameters, some of the times shown will, naturally, fall during daylight, but any difference is too small to affect the appearance of the Moon on that date. Also shown is the **age** of the Moon (the day in the **lunation**), beginning at New Moon, which may be used to determine the best time for observation of specific lunar features.

The Moon

The section on the Moon includes details of any lunar or solar eclipses that may occur during the month (visible from anywhere on Earth). Similar information is given about any important occultations. Mainly, however, this section summarizes when the Moon passes close to planets or the five prominent stars close to the ecliptic. The dates when the Moon is closest to the Earth (at **perigee**) and farthest from it (at **apogee**) are shown in the monthly calendars, and only mentioned here when they are particularly significant, such as the nearest and farthest during the year.

The planets and minor planets

Brief details are given of the location, movement and brightness of the planets from Mercury to Saturn throughout the month. None of the planets can, of course, be seen when they are close to the Sun, so such periods are generally noted. All of the planets may sometimes lie on the opposite side of the Sun to the Earth (at superior conjunction), but

A fireball (with flares approximately as bright as the Full Moon), photographed against a weak auroral display by D. Buczynski from Tarbat Ness in Scotland on January 22, 2017.

in the case of the inferior planets, Mercury and Venus, they may also pass between the Earth and the Sun (at inferior conjunction) and are invisible for a period of time, the length of which varies from conjunction to conjunction. Those two planets are normally easiest to see around either eastern or western elongation, in the evening or morning sky, respectively. Not every elongation is favourable, so although every elongation is listed, only those where observing conditions are favourable are shown in the individual diagrams of events.

The dates at which the superior planets reverse their motion (from direct motion to *retrograde*, and retrograde to direct) and of opposition (when a planet generally reaches its maximum brightness) are given. Some planets, especially distant Saturn, may spend most or all of the year in a single constellation. Jupiter and Saturn are normally easiest to see around opposition, which occurs every year. Mars, by contrast, moves relatively rapidly against the background stars and in some years never comes to opposition.

Uranus is normally magnitude 5.7–5.9, and thus at the limit of naked-eye visibility under exceptionally dark skies, but bright enough to be readily visible in binoculars. Because its orbital period is so long (over 84 years), Uranus moves only slowly along the ecliptic, and often remains within a single constellation for a whole year. The chart on page 25 shows its position during 2024.

Similar considerations apply to Neptune, although it is always fainter (generally magnitude 7.8–8.0), still visible in most binoculars. It takes about 164.8 years to complete one orbit of the Sun. As with Uranus, it frequently spends a complete year in one constellation. Its chart is also on page 25.

In any year, few minor planets ever become bright enough to be detectable in binoculars. Just one, (4) Vesta, on rare occasions brightens sufficiently for it to be visible to the naked eye. Our limit for visibility is magnitude 9.0 and details and charts are given for those objects that exceed that magnitude during the year, normally around opposition. To assist in recognition of a planet or minor planet as it moves against the background stars, the latter are shown to a fainter magnitude than the object at opposition. Minor-planet charts for 2024 are on pages 26 and 27.

The ecliptic charts
Although the ecliptic charts are primarily designed to show the positions and motions of the major planets, they also show the motion of the Sun during the month. The light-tinted area shows the area of the sky that is invisible during daylight, but the darker area gives an indication of which constellations are likely to be visible at some time of the night. The closer a planet is to the border between dark and light, the more difficult it will be to see in the twilight.

The monthly calendar

For each month, a calendar shows details of significant events, including when planets are close to one another in the sky, close to the Moon, or close to any one of five bright stars that are spaced along the ecliptic. The times shown are given in Universal Time (UT), always used by astronomers throughout the year, and which is identical to Greenwich Mean Time (GMT). So during the summer months, they do not show Summer Time, which will always be one hour later than the time shown.

The diagrams of interesting events

Each month, a number of diagrams show the appearance of the sky when certain events take place. However, the exact positions of celestial objects and their separations greatly depend on the observer's position on Earth. When the Moon is one of the objects involved, because it is relatively close to Earth, there may be very significant changes from one location to another. Close approaches between planets or between a planet and a star are less affected by changes of location, which may thus be ignored.

The diagrams showing the appearance of the sky are drawn for the latitude of London, so will be approximately correct for most of Britain and Europe. However, for an observer farther north (say, Edinburgh), a planet or star listed as being north of the Moon will appear even farther north, whereas one south of the Moon will appear closer to it – or may even be hidden (occulted) by it. For an observer farther south than London, there will be corresponding changes in the opposite direction: for a star or planet south of the Moon, the separation will increase, and for one north of the Moon, the separation will decrease. This is particularly important when occultations occur, which may be visible from one location, but not another. However, there are no major occultations visible from Britain at night in 2024.

Ideally, details should be calculated for each individual observer, but this is obviously impractical. In fact, positions and separations are actually calculated for a theoretical observer located at the centre of the Earth.

So the details given regarding the positions of the various bodies should be used as a guide to their location. A similar situation arises with the times that are shown. These are calculated according to certain technical criteria, which need not concern us here. However, they do not necessarily indicate the exact time when two bodies are closest together. Similarly, dates and times are given, even if they fall in daylight, when the objects are likely to be completely invisible. However, such times do give an indication that the objects concerned will be in the same general area of the sky during both the preceding and the following nights.

Data used in this Guide

The data given in this Guide, such as timings and distances between objects, have been computed with a program developed by the US Naval Observatory in Washington DC (MICA, the Multiyear Interactive Computer Almanac), widely regarded as the most accurate computation. As such, the data may differ slightly from information given elsewhere.

Key to the symbols used on the monthy star maps.

Alphard

SEXTANS

ECLIPTIC

E

HYDRA

VIRGO

M67

CANCER

Regulus

LEO

Melotte 111

Denebola

COMA
BERENICES

Algieba

LEO MINOR

M3

Praesepe

M44

Pollux

Castor

CANES VENATICI

URSA
MAJOR

GEMINI

The Plough

BOÖTES

LYNX

M35

CAMELOPARDALIS

Quadrantids

Elnath

URSA MINOR

DRACO

HERCULES

AURIGA

ZENITH

Polaris

N

TAURUS

Capella

The
Kids

LYRA

Vega

CASSIOPEIA

CYGNUS

Caph

Double Cluster

CEPHEUS

Deneb

PERSEUS

Algol

NGC 752

LACERTA

Pleiades

ARIES

TRIANGULUM

Hamal

ANDROMEDA

M31
Andromeda Galaxy

PEGASUS

JUPITER

Alpheratz

PISCES

Great Square
of Pegasus

Mira

CETUS

ECLIPTIC

W

January 01   23:00 GMT
January 15   22:00 GMT
February 01   21:00 GMT

January – Looking North

January – Looking North

Most of the important circumpolar constellations are easy to see in the northern sky at this time of year. *Ursa Major* stands more-or-less vertically above the horizon in the northeast, with the zodiacal constellation of *Leo* rising in the east. To the north, the stars of *Ursa Minor* lie below *Polaris* (the Pole Star). The head of *Draco* is low on the northern horizon, but may be difficult to see unless observing conditions are good. Both *Cepheus* and *Cassiopeia* are readily visible in the northwest, and even the faint constellation of *Camelopardalis* is high enough in the sky for it to be easily visible.

Near the zenith is the constellation of *Auriga* (the Charioteer), with brilliant *Capella* (α Aurigae), directly overhead. Slightly to the west of Capella lies a small triangle of fainter stars, known as 'The Kids'. (Ancient mythological representations of Auriga show him carrying two young goats.) Together with the northernmost bright star in *Taurus*, *Elnath* (β Tauri), the body of Auriga forms a large pentagon on the sky, with The Kids lying on the western side. Farther down towards the west are the constellations of *Perseus* and *Andromeda*, and the Great Square of Pegasus is approaching the horizon.

Meteors

The *Quadrantid* meteor shower actually begins in late December (on 28 December 2023) and continues until January 12. It is one of the strongest and most consistent meteor showers. It comes to maximum on the night of January 3 to 4, around Last Quarter, so moonlight will cause some interference. However, the meteors are bright, bluish- or yellowish-white and may reach a maximum rate of 110 per hour. The parent object is minor planet 2003 EH$_1$.

The shower is named after the former constellation *Quadrans Muralis* (the Mural Quadrant), an early form of astronomical

A late Quadrantid fireball, photographed from Portmahomack, Ross-shire, Scotland on January 15, 2018 at 23:44 UT.

Ursa Major, the key constellation to navigating the northern sky.

instrument. The Quadrantid radiant, marked on the chart, is now within the northernmost part of *Boötes*, roughly halfway between θ Boötis and τ Herculis.

W

PEGASUS

Great Square
of Pegasus

Alpheratz

ANDROMEDA

CASSIOPEIA

M31
Andromeda Galaxy

Double Cluster

NGC 752

PISCES

PERSEUS

TRIANGULUM

Hamal

ARIES

CETUS

Algol

JUPITER

Mira

Pleiades

TAURUS

Hyades

ERIDANUS

The
Kids

Capella

Aldebaran

FORNAX

ZENITH

AURIGA

Elnath

Rigel

CAELUM

CAMELOPARDALIS

M35

LEPUS

S

GEMINI

Betelgeuse

M42
Orion Nebula

ORION

COLUMBA

LYNX

ONI

Alhena

MONOCEROS

Adhara

Castor

CANIS
MINOR

Sirius

Pollux

Procyon

CANIS
MAJOR

URSA MAJOR

LEO MINOR

CANCER

M44
Praesepe

PUPPIS

M67

PYXIS

Algieba

LEO

Melotte 111

Regulus

Alphard

HYDRA

CANES VENATICI

Denebola

SEXTANS

COMA
BERENICES

VIRGO

E

M3

January – Looking South

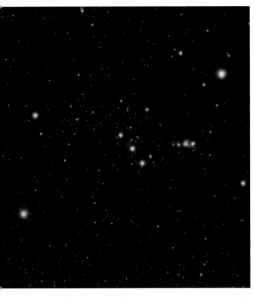

The southern sky is dominated by **Orion**, prominent during the winter months, visible at some time during the night. It is highly distinctive, with a line of three stars that form the 'Belt'. To most observers, the bright star **Betelgeuse** (α Orionis), shows a reddish tinge, in contrast to the brilliant bluish-white **Rigel** (β Orionis). The three stars of the belt lie directly south of the celestial equator. A vertical line of three 'stars' forms the 'Sword' that hangs south of the Belt. With good viewing, the central 'star' appears as a hazy spot, even to the naked eye, and is actually the **Orion Nebula**. Binoculars reveal the four stars of the Trapezium, which illuminate the nebula.

Orion's Belt points up to the northwest towards **Taurus** (the Bull) and orange-tinted **Aldebaran** (α Tauri). Close to Aldebaran, there is a conspicuous 'V' of stars, called the **Hyades** cluster. (Despite appearances, Aldebaran is not part of the cluster.) Farther along, the same line from Orion passes below a bright cluster of stars, the **Pleiades**, or Seven Sisters. Even the smallest pair of binoculars reveals this as a beautiful group of bluish-white stars. The two most conspicuous of the other stars in Taurus lie directly above Orion, and form an elongated triangle with Aldebaran. The northernmost, **Elnath** (β Tauri), was once considered to be part of the constellation of **Auriga**.

The constellation of Orion dominates the sky during this period of the year, and is a useful starting point for recognizing other constellations in the southern sky. Here, orange Betelgeuse, blue-white Rigel and the pinkish Orion Nebula are prominent. Orion can be found in the southern part of the sky (see page 36).

The Moon's phases for January 2024

				Last Quarter		
01	02	03	04 03:30	05	06	07
Day 20	Day 21	Day 22	Day 23	Day 24	Day 25	Day 26
First Quarter						
17	18 03:53	19	20	21	22	23
Day 6	Day 7	Day 8	Day 9	Day 10	Day 11	Day 12

		New Moon				
08	09	10	11 11:57	12	13	14	15	16
Day 27	Day 28	Day 29	Day 30	Day 1	Day 2	Day 3	Day 4	Day 5
Full Moon								
24	25 17:54	26	27	28	29	30	31	
Day 13	Day 14	Day 15	Day 16	Day 17	Day 18	Day 19	Day 20	

January – Moon and Planets

The Earth

The Earth reaches perihelion (the closest point to the Sun in its yearly orbit) on January 3 at 03:39 Universal Time, when it is at a distance of 0.98306994 AU (147,101,081 km).

The Moon

On January 4, at Last Quarter, the Moon is 2°N of **Spica** in **Virgo**. On January 8 it passes 0.8°N of **Antares**, and a few hours later, is 5.7°S of **Venus**. It passes a similar distance south of **Mercury** the next day, and 4.2°S of **Mars** on January 10. On January 14, the waxing crescent Moon is 2.1°S of **Saturn**, and just 1.0°S of **Neptune** the next day. At First Quarter, the Moon is 1.8°N of **Jupiter** and 3.0°N of much fainter **Uranus** the following day. It then passes 9.5°N of **Aldebaran** on January 21.

By January 24, a day before Full Moon, it is 1.7°S of **Pollux**. On January 27 the Moon passes 3.9°N of **Regulus**, between it and **Algieba**.

The planets

Mercury comes to western elongation on January 12 (see the diagram on page 22). **Venus** is bright (mag.-4.0), not far from Mercury in the morning sky. **Mars** (mag.1.3 to 1.4) is in the same area, but lower in the sky. **Jupiter** (mag.-2.4 to -2.5) is in southern **Aries** and **Saturn** (mag.1.0) is in **Aquarius**. **Uranus** (mag.5.7) is in **Aries** and **Neptune** (mag.7.9) in **Pisces**.

The path of the Sun and the planets along the ecliptic in January.

Calendar for January

Early morning 3:00

January 5 • In the early morning, the Moon and Spica are close together, low in the southeast.

Morning 7:00

January 8 • The crescent Moon is in the company of Venus and Antares, with Mercury and Sabik (η Oph) nearby.

Evening 17:00

Saturn
Moon
Fomalhaut
SSW

January 14 • The crescent Moon is near Saturn. Fomalhaut is only six degrees above the horizon.

Evening 23:30

January 18–20 • The Moon passes Jupiter and the Pleiades. Hamal (α Ari), Menkar (α Cet) and Aldebaran are nearby.

Evening 20:30

Castor
Pollux
Moon
Procyon
ESE

January 24 • The full Moon lines up with Pollux and Castor. Procyon is thirty degrees above the horizon.

Evening 20:30

Algieba
Regulus
Moon
E

January 27 • The waning gibbous Moon is due east, between Regulus and Algieba.

February 01   23:00 GMT
February 15   22:00 GMT

February – Looking North

The months of January and February are probably the best time for seeing the section of the Milky Way that runs in the northern and western sky from **Cygnus**, low on the northern horizon, through **Cassiopeia**, **Perseus** and **Auriga** and then down through **Gemini** and **Monoceros** (see chart on page 42). Although not as readily visible as the denser star clouds of the summer Milky Way, on a clear night so many stars may be seen that even a distinctive constellation such as Cassiopeia is not immediately obvious.

The head of **Draco** is now higher in the sky and easier to recognize. **Deneb** (α Cygni), the brightest star in **Cygnus**, may just be visible, almost due north at midnight. **Vega** (α Lyrae) in **Lyra** is so low that it is difficult to see, but may become visible later in the night. The constellation of **Boötes** – sometimes described as shaped like a kite, an ice-cream cone, or the letter 'P' – with orange-tinted **Arcturus** (α Boötis), is beginning to clear the eastern horizon. Arcturus, at magnitude –0.05, is the brightest star in the northern hemisphere of the sky. The inconspicuous constellation of **Coma Berenices** is now well above the horizon in the east. The concentration of faint stars at the northwestern corner somewhat resembles a tiny, detached portion of the Milky Way. This is Melotte 111, an open star cluster (which is sometimes called the Coma Cluster, but must not be confused with the important Coma Cluster of galaxies, Abell 1656, mentioned on page 55).

On the other side of the sky, in the northwest, most of the constellation of **Andromeda** is still easily seen, although **Alpheratz** (α Andromedae), the star that forms the northeastern corner of the Great Square of Pegasus – even though it is actually part of Andromeda – is becoming close to the horizon and more difficult to detect. High overhead, at the zenith, try to make out the very faint constellation of

Lynx. It was introduced in 1687 by the famous astronomer Johannes Hevelius to fill the largely blank area between **Auriga**, **Gemini** and **Ursa Major**, and is reputed to be so named because one needed the eyes of a lynx to detect it.

A very large, and frequently ignored, open star cluster, Melotte 111, also known as the Coma Cluster, is readily visible in the eastern sky during February.

February – Looking South

Apart from **Orion**, the most prominent constellation visible this month is **Gemini**, with its two lines of stars running southwest towards Orion. Many people have difficulty in remembering which is which of the two stars **Castor** and **Pollux**. Castor (α Geminorum), the fainter star (mag. 1.9), is closer to the North Celestial Pole. Pollux (β Geminorum) is the brighter of the two (mag. 1.2), but is farther away from the Pole. Pollux is one of the first-magnitude stars that may be occulted by the Moon. Castor is remarkable because it is actually a multiple system, consisting of no less than six individual stars.

Orion's belt points down to the southeast towards **Sirius**, the brightest star in the sky (at magnitude −1.4) in the constellation of **Canis Major**, the whole of which is now clear of the southern horizon. Forming an equilateral triangle with **Betelgeuse** in Orion and Sirius in Canis Major is **Procyon**, the brightest star in the small constellation of **Canis Minor**. Between Canis Major and Canis Minor is the faint constellation of **Monoceros**, which actually straddles the Milky Way, which, although faint, has many clusters in this area. Directly east of Procyon is the highly distinctive asterism of six stars that form the

The constellation of Gemini. The two brightest stars, visible near the left-hand top corner of the photograph, are Castor (top) and Pollux.

'head' of **Hydra**, the largest of all 88 constellations, and which trails such a long way across the sky that it is only in mid-March around midnight that the whole constellation becomes visible.

The Moon's phases for February 2024

Last Quarter	

01	02 23:18	03	04	05	06	07	08 12:30 Full Moon
Day 21	Day 22	Day 23	Day 24	Day 25	Day 26	Day 27	Day 28

09 22:59 New Moon	10	11	12	13	14	15	16 15:01 First Quarter
Day 29	Day 1	Day 2	Day 3	Day 4	Day 5	Day 6	Day 7

17	18	19	20
Day 8	Day 9	Day 10	Day 11

24 12:30 Full Moon	25	26	27	28	29
Day 15	Day 16	Day 17	Day 18	Day 19	Day 20

(Days 21–14 continue across the chart: 21 Day 12, 22 Day 13, 23 Day 14.)

February – Moon and Planets

The Moon

On February 1 the Moon, just before Last Quarter, is 1.7°N of *Spica* in *Virgo*. By February 5 it is 0.6°N of orange-red *Antares*. One day later it is 5.4°S of brilliant *Venus*. It passes 4.2°S of *Mars* the next day and 3.2°S of *Mercury* later that day. Two days after New Moon (by February 11) the Moon is 1.8°S of *Saturn* and the next day is even closer (0.7°S) to the much fainter *Neptune*. Three days later (February 15) the Moon is 3.2°N of *Jupiter* and the next day the same distance north of *Uranus*. On February 17, one day after First Quarter, the Moon is 9.8°N of *Aldebaran* in *Taurus*. By February 21, it is 1.6°S of *Pollux*, three days before Full Moon. On February 23, the Moon is 3.6°N of *Regulus*, passing between it and *Algieba*.

Occultations

Although there are as many as 14 occultations of *Antares* in 2024 and 9 of *Spica*, none are readily visible from land, visible (if at all) from the oceans. There are no occultations in 2024 of the other bright stars (*Aldebaran*, *Pollux* and *Regulus*) near the ecliptic.

The planets

Mercury is 3.2°N of the Moon on February 8, one day before New Moon. *Venus* remains bright (mag. -3.9) in the morning sky, but is becoming lower towards the horizon. *Mars* is too close to the Sun to be readily visible. *Jupiter* is moving slowly in *Aries* at mag. -2.3 to -2.2. *Saturn* (mag. 1.0) remains in *Aquarius*. *Uranus* (mag. 5.7 to 5.8) is in *Aries* and *Neptune* (mag. 7.9 to 8.0) in *Pisces*.

The path of the Sun and the planets along the ecliptic in February.

Morning 7:00

Moon
Spica

SSW
10°

February 1 • In the morning, the Moon and Spica are close together in the southwestern sky.

Morning 7:00

Sabik •

• Antares

Venus
5

6

SE
SSE
S
10°

February 5–6 • In the morning twilight, the crescent Moon is in the company of Antares and Sabik. Venus is only five degrees above the horizon, due southeast.

Evening 18:00

Pleiades
16

Hamal •

14

Jupiter
15

S
SSW
50°
10°

February 14–16 • The Moon passes between Jupiter and Hamal (α Ari). On February 16, at

Early morning 4:30

Castor •
• Pollux

Moon

W/NW
10°

February 21 • The Moon lines up with Castor and Pollux.

After midnight 0:30

Algieba •

Moon

• Regulus

S
50°
10°

February 24 • High in the south, the Full Moon passes

SSW

VIRGO

SERPENS
CAPUT

Arcturus

Alphecca

HERCULES

COMA
BERENICES

M3

BOÖTES

CORONA
BOREALIS

M13

Melotte 111

CANES VENATICI

Keystone

LYRA

Vega

The Plough

DRACO

URSA MAJOR

CYGNUS

URSA MINOR

CEPHEUS

ZENITH

Deneb

μ
Garnet Star

CAMELOPARDALIS

Polaris

Errai

N

δ

CASSIOPEIA

Caph

LACERTA

ster

Galaxy

March – Looking North

The Sun crosses the celestial equator on Wednesday, March 20, at the vernal equinox, when day and night are of almost equal length, and the northern season of spring is considered to begin. (The hours of daylight and darkness change most rapidly around the equinoxes in March and September.) It is also in March that Daylight Saving Time (DST) begins (on Saturday–Sunday, March 30–31) so the charts show the appearance at 11 p.m. for March 1 and 10 p.m. (BST) for April 1. (In Europe, Daylight Saving Time is introduced on the same date.)

Early in the month, the constellation of **Cepheus** lies almost due north, with the distinctive 'W' of **Cassiopeia** to its west. Cepheus lies across the border of the Milky Way and is often described as like the gable-end of a house or a church tower and steeple. Despite the large number of stars revealed at the base of the constellation by binoculars, one star stands out because of its deep red colour. This is **μ Cephei**, also known as the Garnet Star, because of its striking colour. It is a truly gigantic star, a red supergiant, and one of the largest stars known. It is about 1,400 times the diameter of the Sun, and if placed in the Solar System would extend beyond the orbit of Jupiter. (Betelgeuse, in Orion, is also a red supergiant, but it is 'only' about 500 times the diameter of the Sun.)

Another famous, and very important star in Cepheus is **δ Cephei**, which is the prototype for the class of variable stars known as Cepheids. These giant stars show a regular variation in their luminosity, and there is a direct relationship between the period of the changes in magnitude and the stars' actual luminosity. From a knowledge of the period of any Cepheid, its actual luminosity – known as its absolute magnitude – may be derived. A comparison of its apparent magnitude on the sky and its absolute magnitude enables the star's exact distance to be determined. Once the distances to the first Cepheid variables had been established,

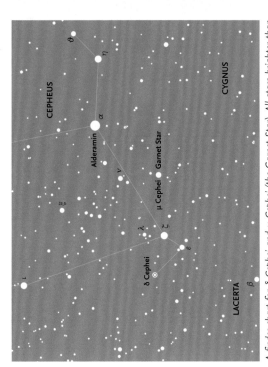

A finder chart for δ Cephei and μ Cephei (the Garnet Star). All stars brighter than magnitude 7.5 are shown. A photograph of the area is on page 49.

examples in more distant galaxies provided information about the scale of the universe. Cepheid variables are the first major 'rung' in the cosmic distance ladder. Both important stars are shown on the accompanying chart and in the photograph on page 49.

Below Cepheus to the east (to the right), it may be possible to catch a glimpse of **Deneb** (α Cygni), just above the horizon. Slightly farther round towards the northeast, **Vega** (α Lyrae) is marginally higher in the sky. From southern Britain, Deneb is just far enough north to be circumpolar (although difficult to see in January and February because it is so low on the northern horizon). Vega, by contrast, farther south, is completely hidden during the depths of winter.

March 01   23:00 GMT
March 15   22:00 GMT
April 01   22:00 BST (GMT + 1hr)

March: Looking South

TRIANGULUM
Hama
ARIES
JUPITER
PERSEUS
Algol
TAURUS
ERIDANUS
Pleiades
Hyades
Aldebaran
ORION
Elnath
Rigel
The Kids
M42
Orion Nebula
Capella
M35
Betelgeuse
LEPUS
AURIGA
Alhena
Sirius
GEMINI
CANIS
MINOR
MONOCEROS
CANIS
MAJOR
CAMELOPARDALIS
LYNX
Castor
Pollux
Procyon
Adhara
CANCER
PUPPIS
M44
Praesepe
M67
ZENITH
Alphard
PYXIS
LEO MINOR
Algieba
HYDRA
S
URSA MAJOR
Regulus
SEXTANS
ANTLIA
DRACO
LEO
The Plough
Ghost of Jupiter
CANES VENATICI
Melotte 111
CRATER
M3
COMA BERENICES
Denebola
BOÖTES
VIRGO
CORVUS
ECLIPTIC
M13
Arcturus
Spica
HERCULES
Alphecca
CORONA
BOREALIS
SERPENS
CAPUT
E

March – Looking South

Due south at 22:00 at the beginning of the month, lying between the constellations of **Gemini** in the west and **Leo** in the east, and fairly high in the sky above the head of **Hydra**, is the faint, and rather undistinguished zodiacal constellation of **Cancer**. Rather like the triskelion, the symbol for the Isle of Man, it has three 'legs' radiating from the centre, where there is an open cluster, M44 or **Praesepe** ('the Manger' but also known as 'the Beehive'). On a clear night this cluster, known since antiquity, is just a hazy spot to the naked eye, but appears in binoculars as a group of dozens of individual stars.

Also prominent in March is the constellation of Leo, with the 'backward question mark' (or 'Sickle') of bright stars forming the head of the mythological lion. **Regulus** (α Leonis) – the 'dot' of the 'question mark' or the handle of the sickle and the brightest star in Leo – lies very close to the ecliptic and is one of the few first-magnitude stars that may be occulted by the Moon. However, there are no occultations of Regulus in 2024, nor of any of the other four bright stars near the ecliptic.

The Moon's phases for March 2024

First Quarter

01	02	03 15:23	04	05	06	07	08	09	10 09:00
Day 21	Day 22	Day 23	Day 24	Day 25	Day 26	Day 27	Day 28	Day 29	Day 30

Last Quarter ... New Moon

11	12	13	14	15	16
Day 1	Day 2	Day 3	Day 4	Day 5	Day 6

17 04:11	18	19	20	21	22	23	24	25 07:00	26
Day 7	Day 8	Day 9	Day 10	Day 11	Day 12	Day 13	Day 14	Day 15	Day 16

Full Moon

27	28	29	30	31
Day 17	Day 18	Day 19	Day 20	Day 21

The constellation of Cepheus, with the 'W' of Cassiopeia on the left. The locations of both δ and μ Cephei are shown on the chart on page 47.

March – Moon and Planets

The Moon

On March 3, at Last Quarter, the Moon is 0.4°N of *Antares*. On March 8 it passes 3.5°S of *Mars*. Later the same day it is 3.3°S of *Venus*. The next day (March 9) it is 1.5°S of *Saturn* (mag. 1.0). On March 11, a few hours after New Moon, it is 0.5°S of *Neptune* (mag. 8.0) and the next day it is 1.0°S of *Mercury* (mag. -1.3). On March 14 it is 4.0°N of *Jupiter* (mag. -2.1) and later that day 3.4°N of *Uranus* (mag. 5.8). On March 15 it is 9.9°S of *Aldebaran* in *Taurus*. On March 19, just after First Quarter, it is 1.4°S of *Pollux*. By March 22 it is 3.6°N of *Regulus*. At Full Moon (March 25) there is a penumbral lunar eclipse (visible from the USA) when the Moon may just touch the umbra. The next day the Moon is 1.4°N of *Spica*. On March 30 the Moon is 0.3°N of *Antares*.

The Planets

Mercury (mag. -0.2) reaches greatest eastern elongation on March 24 (see the diagram on page 22). *Venus* is bright (mag. -3.8) but is closing rapidly on the Sun, and may be glimpsed early in the month. *Mars* (mag. 1.3 to 1.2) is similarly placed. *Jupiter* (mag. -2.2 to -2.1) is in *Aries* and *Saturn* (mag. 1.0 to 1.1) remains in *Aquarius*. *Uranus* (mag. 5.8) is in *Aries* and *Neptune* (mag. 8.0) in *Pisces*.

The path of the Sun and the planets along the ecliptic in March.

Calendar for March

03	08:54	Antares 0.4°S of the Moon
03	15:23	Last Quarter
03	18:01	Minor planet (3) Juno at opposition (mag. 8.7)
08	05:00	Mars (mag. 1.2) 3.5°N of the Moon
08	08:18 *	Neptune (mag. 8.0) 0.5°S of Mercury (mag. -1.5)
08	17:00	Venus (mag. -3.8) 3.3°N of the Moon
09	17:28	Saturn (mag. 1.0) 1.5°N of the Moon
10	07:04	Moon at perigee = 356,896 km (closest of the year)
10	19:26	Neptune (mag. 8.0) 0.5°N of the Moon
11	02:31	Mercury (mag. -1.3) 1.0°N of the Moon
11	09:00	New Moon
14	01:02	Jupiter (mag. -2.1) 4.0°S of the Moon
14	11:35	Uranus (mag. 5.8) 3.4°S of the Moon
15	23:43	Aldebaran 9.9°N of the Moon
17	04:11	First Quarter
19	07:24	Pollux 1.4°N of the Moon
22	05:27	Regulus 3.6°S of the Moon
23	15:45	Moon at apogee = 406,294 km
24	22:34	Mercury at greatest elongation (18.7°E, mag. -0.2)
25	07:00	Full Moon
25	07:14	Penumbral lunar eclipse
26	20:23	Spica 1.4°S of the Moon
30	15:03	Antares 0.3°S of the Moon
31		Summer Time begins (BST in UK)

* These objects are close together for an extended period around this time.

Morning 6:00

March 3 • The Last Quarter Moon is close to Antares, low in the southern sky. Sabik is close by.

Evening 20:30

March 13–16 • In the west, the waxing crescent Moon passes Jupiter, Hamal, the Pleiades and Aldebaran. It is almost at First Quarter when it reaches Elnath (β Tau).

Evening 20:00

March 20–21 • High in the southeast, the waxing gibbous Moon passes between Regulus and Algieba.

Evening 21:30

March 26 • Shortly after Full Moon, it is close to Spica, low in the southeast.

Morning 5:00 (BST)

March 30–31 • Low in the south, the waning gibbous Moon passes Antares again, with Sabik nearby.

April – Looking North

00:00 BST (GMT + 1hr)
23:00 BST (GMT + 1hr)
22:00 BST (GMT + 1hr)

April 01
April 15
May 01

April – Looking North

Cygnus and the brighter regions of the Milky Way are now becoming visible, running more-or-less parallel with the horizon in the early part of the night. Rising in the northeast is the small constellation of **Lyra** and the distinctive 'Keystone' of **Hercules** above it. This asterism is very useful for locating the bright globular cluster M13 (see map on page 59), which lies on one side of the quadrilateral. The winding constellation of **Draco** weaves its way from the four stars that mark its 'head', on the border with Hercules, to end at **Giausar** (λ Draconis) between **Polaris** (α Ursae Minoris) and the 'Pointers', **Dubhe** and **Merak** (α and β Ursae Majoris, respectively). **Ursa Major** is 'upside down' high overhead, near the zenith. The constellation of **Gemini** stands almost vertically in the west. **Auriga** is still clearly seen in the northwest, but, by the end of the month, the southern portion of **Perseus** is starting to dip below the northern horizon. The very faint constellation of **Camelopardalis** lies in the northwest between Polaris and the constellations of Auriga and Perseus.

Meteors

There is one moderate meteor shower this month, the **Lyrids** (often known as the **April Lyrids**). This year the shower begins on April 14, one day before First Quarter, but peaks on April 22 to 23 at Full Moon, so begins reasonably favourably but becomes very unfavourable. There is another, stronger shower, the **η-Aquariids**, that begins on April 19, when the Moon is waxing gibbous, and continues until May 28, well after the next New Moon. The shower is best seen in early May.

The constellation of Hercules (centre), with Lyra (bottom left) and Corona Borealis (top right).

April – Looking South

April 01   00:00 BST (GMT +1hr)
April 15   23:00 BST (GMT +1hr)

PERSEUS
Pleiades
The Kids
Elnath
TAURUS
ORION
Capella
AURIGA
Betelgeuse
M35
CANIS MINOR
MONOCEROS
GEMINI
Alhena
LYNX
Castor
Pollux
CANCER
Procyon
M44
Praesepe
M67
PUPPIS
Alphard
HYDRA
PYXIS
LEO MINOR
Algieba
LEO
Regulus
SEXTANS
Ghost of Jupiter
ANTLIA
URSA MAJOR
ECLIPTIC
ZENITH
The Plough
CRATER
CANES VENATICI
Melotte 111
COMA BERENICES
Denebola
CORVUS
DRACO
M3
Spica
CENTAURUS
BOÖTES
VIRGO
Arcturus
HYDRA
Alphecca
CORONA BOREALIS
LIBRA
M13
Keystone
SERPENS CAPUT
HERCULES
Vega
LYRA
OPHIUCHUS
CYGNUS
M57
Ring Nebula
April Lyrids

S

E

April – Looking South

Leo is the most prominent constellation in the southern sky in April, and vaguely looks like the creature after which it is named. *Gemini*, with *Castor* and *Pollux*, remains clearly visible in the west, and *Cancer* lies between the two constellations. To the east of Leo, the whole of *Virgo*, with *Spica* (α Virginis) its brightest star, is well clear of the horizon. Below Leo and Virgo, the complete length of *Hydra* is visible, running beneath both constellations, with *Alphard* (α Hydrae) halfway between *Regulus* and the southwestern horizon. Farther east, the two small constellations of *Crater* and the rather brighter *Corvus* lie between Hydra and Virgo.

Boötes and *Arcturus* are prominent in the eastern sky, together with the circlet of *Corona Borealis*, framed by Boötes and the neighbouring constellation of *Hercules*. Between Leo and Boötes lies the constellation of *Coma Berenices*, notable for being the location of the open cluster Melotte 111 (see page 41) and the Coma Cluster of galaxies (Abell 1656). There are about 1000 galaxies in this cluster, which is located near the North Galactic Pole, where we are looking out of the plane of the Galaxy and are thus able to see deep into space. Only about ten of the brightest galaxies in the Coma Cluster are visible with the largest amateur telescopes.

The distinctive constellation of Leo, with Regulus and 'The Sickle' on the west. Algieba (γ Leonis), north of Regulus, appearing double, is a multiple system of four stars.

The Moon's phases for April 2024

		Last Quarter						
01	02 03:15	03	04	05	06	07	08 18:21	
Day 22	Day 23	Day 24	Day 25	Day 26	Day 27	Day 28	Day 29	

New Moon

					First Quarter	
09	10	11	12	13	14	15 19:13	16
Day 1	Day 2	Day 3	Day 4	Day 5	Day 6	Day 7	Day 8

17	18	19	20	21	22	23 23:49	24
Day 9	Day 10	Day 11	Day 12	Day 13	Day 14	Day 15	Day 16

Full Moon

25	26	27	28	29	30
Day 17	Day 18	Day 19	Day 20	Day 21	Day 22

April – Moon and Planets

The Moon

On April 8, at New Moon, there is a total solar eclipse, with the track covering Mexico, the eastern states of the USA and eastern Canada (see pages 20 and 21). The Moon then passes the **Pleiades** and by April 12 is 10°N of **Aldebaran** in **Taurus**. It is 1.5°S of **Pollux** by April 15 and 3.5°N of **Regulus**, between it and **Algieba** on April 18. Before Full Moon on April 23, it is 1.4°N of **Spica** in **Virgo** and 0.3°N of **Antares** by April 26.

The planets

Mercury is too close to the Sun to be visible this month. **Venus** is similarly placed. **Mars** is moving towards the Sun, but may be glimpsed in the morning sky early in the month. **Jupiter** (mag. -2.0) is in **Aries**, crossing into **Taurus** at the end of the month. **Saturn** (mag. 1.1 to 1.2) is moving slowly in **Aquarius**. **Uranus** (mag. 5.8) remains in **Aries** and **Neptune** (mag. 8.0 to 7.9) in **Pisces**.

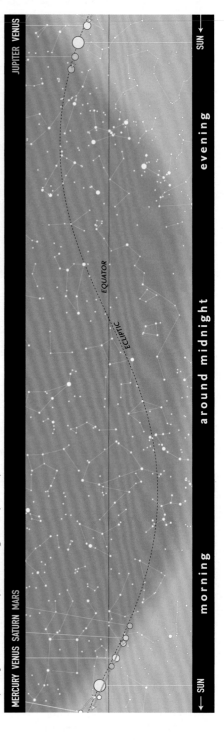

The path of the Sun and the planets along the ecliptic in April.

Evening 21:00 (BST)

Aldebaran

Algol

Pleiades 11

Jupiter 10

Hamal

W

WNW

April 10–11 · *After sunset, the narrow crescent Moon passes Jupiter and the Pleiades, surrounded by Aldebaran, Algol (β Per) and Hamal (α Ari).*

Early morning 3:00 (BST)

Denebola

Algieba

19

18

Regulus

W

WNW

April 18–19 · *The waxing gibbous Moon passes between Regulus and Albiebra, in the western sky. Denebola (β Leo) is nearby.*

Evening 21:00 (BST)

Aldebaran

Pleiades

Jupiter

WNW

April 27 · *After sunset, Jupiter is eight degrees below the Pleiades.*

Early morning 4:00 (BST)

Moon

Spica

SW

WSW

April 23 · *The almost Full Moon is close to Spica, in the southwest.*

Early morning 4:00 (BST)

Sabik

Antares

26

27

S

SSW

April 26–27 · *The waning gibbous Moon passes Antares, near the south.*

May, Looking North

May 01   00:00 BST (GMT + 1hr)
May 15   23:00 BST (GMT + 1hr)

May – Looking North

Cassiopeia is now low over the northern horizon and, to its west, the southern portions of both *Perseus* and *Auriga* are becoming difficult to observe, although the *Double Cluster*, between Perseus and Cassiopeia is still clearly visible. The *Andromeda Galaxy* is now too low to be readily visible. The constellations of *Lyra, Cepheus, Ursa Minor* and the whole of *Draco* are well placed in the sky. *Gemini*, with *Castor* and *Pollux*, is sinking towards the western horizon. *Capella* (α Aurigae) and the asterism of *The Kids* are still clear of the horizon.

In the east, two of the stars of the 'Summer Triangle,' *Vega* (α Lyrae) and *Deneb* (α Cygni), are clearly visible, and the third star, *Altair* in *Aquila*, is beginning to climb above the horizon. The whole of *Cygnus* is now visible. The sprawling constellation of *Hercules* is high in the east and the brightest globular cluster in the northern hemisphere, M13, is visible to the naked eye on the western side of the asterism known as the *Keystone.*

Three faint constellations may be identified before the lighter nights of summer make them difficult objects. Below Cepheus, in the northeastern sky is the zig-zag constellation of *Lacerta*, while to the west, above Perseus and Auriga is *Camelopardalis* and, farther west, the line of faint stars forming *Lynx.*

Later in the night (and in the month) the westernmost stars of *Pegasus* begin to come into view, while the stars of *Andromeda* are skimming the northeastern horizon and the Andromeda Galaxy may become visible. High overhead, *Alkaid* (η Ursae Majoris), the last star in the 'tail' of the Great Bear, is close to the zenith, while the main body of the constellation has swung round into the western sky.

Meteors

The *η-Aquariids* are one of the two meteor showers associated with Comet 1P/Halley (the other being the Orionids, in October). The η-Aquariids are not particularly favourably placed for northern-

hemisphere observers, because the radiant is near the celestial equator, near the 'Water Jar' in Aquarius, well below the horizon until late in the night (around dawn). However, meteors may still be seen in the eastern sky even when the radiant is below the horizon. There is a radiant map for the η-Aquariids on page 30.

Their maximum in 2024, on May 6, occurs two days before New Moon, so is very favourable. Maximum hourly rate is about 50–55 per hour and a large proportion (about 25 per cent) of the meteors leave persistent trains.

CORONA BOREALIS

HERCULES

A finder chart for M13, the finest globular cluster in the northern sky. All stars down to magnitude 7.5 are shown.

May, Looking South

00:00 BST (GMT + 1hr)   May 01
23:00 BST (GMT + 1hr)   May 15
22:00 BST (GMT + 1hr)   June 01

May – Looking South

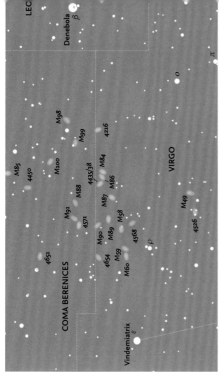

Early in the night, the constellation of *Virgo*, with *Spica* (α Virginis), lies due south, with *Leo* and both *Regulus* and *Denebola* (α and β Leonis, respectively) to its west still well clear of the horizon. Later in the night, the rather faint zodiacal constellation of *Libra* becomes visible and, to its east, the ruddy star *Antares* (α Scorpii) begins to climb up over the horizon.

Virgo contains the nearest large cluster of galaxies, which is the centre of the Local Supercluster, of which the Milky Way galaxy forms part. The Virgo Cluster contains some 2000 galaxies, the brightest of which are visible in amateur telescopes.

Arcturus in *Boötes* is high in the south, with the distinctive circlet of *Corona Borealis* clearly visible to its east. The brightest star (α Coronae Borealis) is known as *Alphecca*. The large constellation of *Ophiuchus* (which actually crosses the ecliptic, and is thus the 'thirteenth' zodiacal constellation) is climbing into the eastern sky. Before the constellation boundaries were formally adopted by the International Astronomical Union in 1930, the southern region of Ophiuchus was regarded as

A finder chart for some of the brightest galaxies in the Virgo Cluster (see page 60). All stars brighter than magnitude 8.5 are shown.

forming part of the constellation of *Scorpius*, which had been part of the zodiac since antiquity.

The Moon's phases for May 2024

May – Moon and Planets

The Moon

On May 3, the Moon is 0.8°S of **Saturn** (mag. 1.2) and the next day is 0.3°S of much fainter **Neptune** (mag. 7.9). By May 5 it is 0.2°N of **Mars** (mag. 1.0). On May 6 it is 3.8°N of **Mercury** (mag. 0.6) and the next day 3.5°N of much brighter **Venus** (mag. -3.9). At New Moon on May 8, it is 3.6°N of faint **Uranus** (mag. 5.8) and a few hour later 4.3°N of **Jupiter** (mag. -2.0). On May 9 it is 9.9°N of **Aldebaran** in **Taurus**. On May 12 the Moon is 1.6°S of **Pollux**. At First Quarter on May 15, it is 3.5°N of **Regulus**. By May 20, the Moon is 1.4°N of **Spica** in **Virgo**. On May 24, the Moon (just past Full) is 0.4°N of **Antares**. By May 31, the Moon is 0.4°S of **Saturn**.

The planets

Mercury (mag. 0.4) reaches greatest elongation west in the evening sky on May 9. By May 31 (at mag. -0.7) it is 1.4°S of **Uranus** (mag. 5.8). **Venus** passes conjunction with the Sun, so is invisible this month. **Mars** is moving towards the Sun, so is also invisible. **Jupiter** (mag. -2.0), now in **Taurus**, is moving towards the **Hyades**. **Saturn** is moving slowly east in **Aquarius**. **Uranus** (mag. 5.8) remains in **Aries** and **Neptune** (mag. 7.9) in **Pisces**.

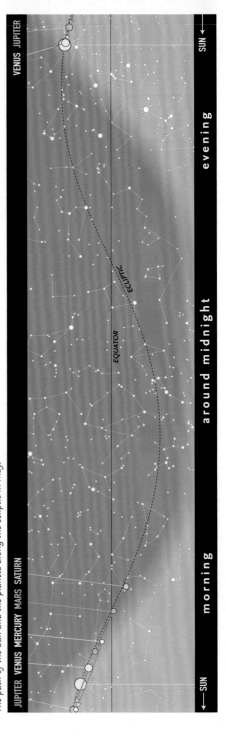

The path of the Sun and the planets along the ecliptic in May.

Calendar for May

01	11:27	Last Quarter
03	22:32	Saturn (mag. 1.2) 0.8°N of the Moon
04	18:55	Neptune (mag.-7.9) 0.3°N of the Moon
05	02:25	Mars (mag. 1.0) 0.2°S of the Moon
05	22:04	Moon at perigee = 363,163 km
06		η-Aquariid meteor shower maximum
06	08:25	Mercury (mag. 0.6) 3.8°S of the Moon
07	16:03	Venus (mag. -3.9) 3.5°S of the Moon
08	03:22	New Moon
08	12:51	Uranus (mag. 5.8) 3.6°S of the Moon
08	18:14	Jupiter (mag. -2.0) 4.3°S of the Moon
09	18:54	Aldebaran 9.9°S of the Moon
09	21:29	Mercury at greatest elongation (26.4°W, mag. 0.4)
12	22:54	Pollux 1.6°N of the Moon
15	11:48	First Quarter
15	19:24	Regulus 3.5°S of the Moon
17	18:59	Moon at apogee = 404,640 km
19	15:06	Minor planet (2) Pallas at opposition (mag. 9.0)
20	10:03	Spica 1.4°S of the Moon
23	13:53	Full Moon
24	03:10	Antares 0.4°S of the Moon
30	17:13	Last Quarter
31	01:00 *	Uranus (mag. 5.8) 1.4°N of Mercury (mag. -0.7)
31	08:09	Saturn (mag. 1.2) 0.4°N of the Moon

* These objects are close together for an extended period around this time.

Morning 5:00 (BST)

May 3–4 • In the morning twilight, the crescent Moon passes below Saturn (mag. 1.1). Mars is equally bright, but closer to the horizon. It will probably be lost in twilight.

Early morning 3:10 (BST)

May 24 • In the early morning, the Full Moon is close to Antares, low in the south-southwest. Sabik is about fifteen degrees higher.

Evening 22:00 (BST)

May 15 • Just past First Quarter, the Moon is between Regulus and Algieba. Denebola (β Leo) is farther south.

Early morning 4:00 (BST)

May 31 • The Last Quarter Moon is close to Saturn (mag. 1.0). Mars is equally bright and almost due east.

June – Looking North

June 01   00:00 BST (GMT + 1hr)
June 15   23:00 BST (GMT + 1hr)

June – Looking North

Around summer solstice (June 20) even in southern England and Ireland a form of twilight persists throughout the night. Farther north, in Scotland, the sky remains so light that most of the fainter stars and constellations are invisible. There, even brighter stars, such as the seven stars making up the well-known asterism known as the *Plough* in *Ursa Major* may be difficult to detect except around local midnight, 00:00 UT (01:00 BST).

But there is one compensation during these light nights: even southern observers may be lucky enough to witness a display of noctilucent clouds (NLC). These are highly distinctive clouds shining with an electric-blue tint, observed in the sky in the direction of the North Pole. They are the highest clouds in the atmosphere, occurring at altitudes of 80–85 km, far above all other clouds. They are only visible during summer nights, for about a month or six weeks on either side of the solstice, when observers are in darkness, but the clouds themselves remain illuminated by sunlight, reaching them from the Sun, itself hidden below the northern horizon.

Noctilucent clouds, photographed by Alan Tough from Nairn in Scotland, on May 31, 2020, at 00:28.

June – Looking South

June 01   00:00 BST (GMT + 1hr)
June 15   23:00 BST (GMT + 1hr)
July 01   22:00 BST (GMT + 1hr)

June – Looking South

Although the persistent twilight makes observing even the southern sky difficult, the rather undistinguished constellation of **Libra** lies almost due south. The red supergiant star **Antares** – the name means the 'Rival of Mars' – in **Scorpius** is visible slightly to the east of the meridian, but the 'tail' or 'sting' remains below the horizon. Higher in the sky is the large constellation of **Ophiuchus** (the 'Serpent Bearer'), lying between the two halves of the constellation of **Serpens: Serpens Caput** ('Head of the Serpent') to the west and **Serpens Cauda** ('Tail of the Serpent') to the east. (Serpens is the only constellation to be divided into two distinct parts.) The ecliptic runs across Ophiuchus, and the Sun actually spends far more time in the constellation than it does in the 'classical' zodiacal constellation of Scorpius, a small area of which lies between Libra and Ophiuchus.

Higher in the southern sky, the three constellations of **Boötes**, **Corona Borealis** and **Hercules** are now better placed for observation than at any other time of the year. This is an ideal time to observe the fine globular cluster of M13 in Hercules (see finder chart on page 59).

The Moon's phases for June 2024

The constellations of Boötes and Corona Borealis are high in the sky in June. Arcturus has an orange tint and is the brightest star (mag. -0.05) in the northern celestial hemisphere.

June – Moon and Planets

The Moon

On June 1, the Moon passes extremely close and to the south of **Neptune**, but there is no occultation. The next day it is 2.4°N of **Mars**. On June 5, the day before New Moon, it passes 3.7°N of **Uranus** (mag. 5.8) and later that day 4.7°N of **Jupiter** (mag. -2.0). Later again, that same day, it is 4.7°N of **Mercury** (mag. -1.2). It is 9.9°N of **Aldebaran** on June 6 (at New Moon). By June 9 it is 1.7°S of **Pollux** and by June 12, 3.3°N of **Regulus** in **Leo**. On June 16 it is 1.2°N of **Spica**. By June 20, the Moon is 0.3°N of **Antares**. By June 27, the Moon is 0.1°N of **Saturn** and, the next day (June 28) it is 0.3°N of **Neptune** (mag. 7.9).

The planets

Mercury moves from the morning to the evening sky and is mag. -1.2 and 4.7°S of the Moon on June 5 (one day before New Moon). **Venus** is very close to the Sun in **Gemini**. Mars (mag. 1.0) moves from **Pisces** into **Aries**. Jupiter (mag. -2.0) is in **Taurus**, 4.7°S of the Moon on June 5 (one day before New Moon). **Saturn** (mag. 1.1) is moving slowly east in **Aquarius. Uranus** (mag. 5.8) is now in **Taurus** and **Neptune** (mag. 7.9) remains in **Pisces**.

The path of the Sun and the planets along the ecliptic in June.

Calendar for June

Early morning 4:00 (BST)

Evening 22:00 (BST)

June 2–3 • Shortly before sunrise, the crescent Moon passes Mars. Hamal is nearby.

June 8–9 • In the evening dawn, the narrow crescent Moon passes below the Twin Stars, Pollux and Castor.

Evening 22:00 (BST)

Evening 23:30 (BST)

Early morning 4:00 (BST)

June 16 • The Moon and Spica are side by side.

June 19–20 • The nearly Full Moon passes below Antares. Sabik is nearby.

June 27–28 • The Moon passes Saturn. Fomalhaut and Diphda (β Cet) are close to the horizon.

July – Looking North

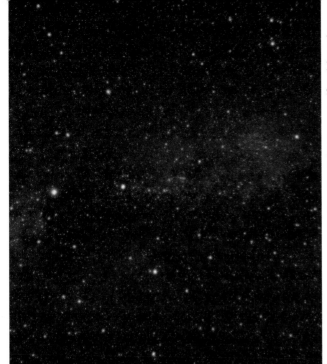

As in June, light nights and the chance of observing noctilucent clouds persist throughout July, but later in the month (and particularly after midnight) some of the major constellations begin to be more easily seen. **Capella,** the brightest star in **Auriga** (most of which is too low to be visible), is skimming the northern horizon. **Cassiopeia** is clearly visible in the northeast and **Perseus,** to its south, is beginning to climb clear of the horizon. The band of the Milky Way, from Perseus through Cassiopeia towards **Cygnus,** stretches up into the northeastern sky. If the sky is dark and clear, you may be able to make out the small, faint constellation of **Lacerta,** lying across the Milky Way between Cassiopeia and Cygnus. In the east, the stars of **Pegasus** are now well clear of the horizon, with the main line of stars forming **Andromeda** roughly parallel to the horizon in the northeast. **Alpheratz** (α Andromedae) is actually the star at the northeastern corner of the **Great Square of Pegasus. Cepheus** and **Ursa Major** are on opposite sides of **Polaris** and **Ursa Minor,** in the east and west, respectively. The head of **Draco** is very close to the zenith so the whole of this winding constellation is readily seen.

Meteors

July brings increasing meteor activity, mainly because there are several minor radiants active in the constellations of **Capricornus** and **Aquarius.** Because of their location, however, observing conditions are not particularly favourable for northern-hemisphere observers, although the first shower, the **α-Capricornids,** active from July 3 to August 15 (peaking July 30, with a tail to August 15), does often produce very bright fireballs. The maximum rate, however, is only about 5 per hour. The parent body is Comet 169P/NEAT. The most prominent shower is probably that of the **Southern δ-Aquariids,** which are active from around July 12 to August 23, with a peak on July 30, although even then the rate is unlikely to reach 25 meteors per hour. In this case,

Cygnus, sometimes known as the 'Northern Cross', depicts a swan 'flying' down the Milky Way towards Sagittarius. The brightest star, Deneb (α Cygni), represents the tail, and Albireo (β Cygni) marks the position of the head, and lies near the bottom of the image.

the parent body may be Comet 96P/Machholz. This year, both shower maxima occur when the Moon is a few days before New Moon so observing conditions are reasonably favourable. A chart showing the **δ-Aquariids** radiant is shown on page 30. The **Perseids** begin on July 17 and peak on August 12–13.

July – Looking South

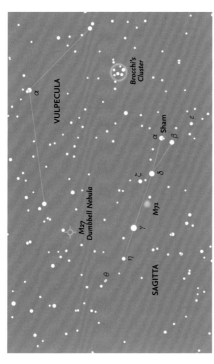

Although part of the constellation remains hidden, this is perhaps the best time of year to see **Scorpius**, with deep red **Antares** (α Scorpii), glowing just above the southern horizon. At around midnight (UT), 01:00 BST, part of **Sagittarius**, with the distinctive asterism of the 'Teapot', and the dense star clouds of the centre of the Milky Way, are just visible in the south. The **Great Rift** – actually dust clouds that hide the more distant stars – runs down the Milky Way from **Cygnus** towards Sagittarius. Towards its northen end is the small constellation of **Sagitta** and the planetary nebula **M27** (the Dumbbell Nebula) in the otherwise insignificant constellation of **Vulpecula**. The sprawling constellation of **Ophiuchus** lies close to the meridian for a large part of the month, separating the two halves of the constellation of **Serpens**. The western half is called **Serpens Caput** (Head of the Serpent) and the eastern part **Serpens Cauda** (Tail of the Serpent). In the east, the bright 'Summer Triangle', consisting of **Vega** in **Lyra**, **Deneb** in Cygnus and **Altair** in **Aquila**, begins to dominate the southern sky, as it will throughout August and into September. The small constellation of Lyra, with Vega and a distinctive quadrilateral of stars to its east and south, lies not far south of the zenith.

A finder chart for M27, the Dumbbell Nebula, a relatively bright (magnitude 8) planetary nebula – a shell of material ejected in the late stages of a star's lifetime – in the constellation of Vulpecula. All stars brighter than magnitude 7.5 are shown.

The Moon's phases for July 2024

New Moon — 05 22:57 (Day 29)

Full Moon — 21 10:17 (Day 15)

First Quarter — 13 22:49

Last Quarter — 28 02:51

01 Day 25	02 Day 26	03 Day 27	04 Day 28	05 22:57 Day 29	06 Day 30	07 Day 1	08 Day 2
09 Day 3	10 Day 4	11 Day 5	12 Day 6	13 22:49 Day 7	14 Day 8	15 Day 9	16 Day 10
17 Day 11	18 Day 12	19 Day 13	20 Day 14	21 10:17 Day 15	22 Day 16	23 Day 17	24 Day 18
25 Day 19	26 Day 20	27 Day 21	28 02:51 Day 22	29 Day 23	30 Day 24	31 Day 25	

July – Moon and Planets

The Moon

On July 1, the Moon is 4.1°N of **Mars** (mag.1.0). The next day it is 4.0°N of **Uranus** (mag.5.8). By July 3 it is 5.0°N of **Jupiter** (mag.-2.0) and, later, 9.9°N of **Aldebaran**. On July 6 the Moon is 3.9°N of **Venus** (mag.-3.9) and later 1.8°S of **Pollux** in **Gemini**. The next day the Moon is 3.2°N of **Mercury** (mag.-0.3). By July 9 it passes 3.0°N of **Regulus**, between it and **Algieba**. On July 14, one day after First Quarter, it is 0.9°N of **Spica**. On July 17 the Moon is 0.2°N of **Antares** in **Scorpius**. By July 24 the waning gibbous Moon is 0.4°N of **Saturn** (mag.0.9) and the next day 0.6°N of **Neptune** (mag.7.8). On July 29, one day after Last Quarter, the waning crescent Moon is 4.2°N of **Uranus** (mag.5.8). By July 30 it passes 10.1°N of **Aldebaran**, 5.0°N of **Mars** (mag.0.9) and 5.4°N of **Jupiter** (mag.-2.1).

The planets

Mercury is moving eastwards past the Sun, and comes to eastern elongation on July 22 at mag.0.3. **Venus** is bright (mag.-3.9) in the evening sky. **Mars** (mag.1.0 to 0.9) is in **Aries**, moving into **Taurus** by the end of the month. **Jupiter** (mag.-2.0 to -2.1) is moving slowly eastwards in **Taurus**. **Saturn** (mag.-2.0 to 0.9) is in **Aquarius** and begins retrograde motion on July 4. **Uranus** (mag.5.8) is in **Taurus** and **Neptune** (mag.7.9) remains in **Pisces**.

VENUS   SUN

VENUS MERCURY

JUPITER   MARS   SATURN

ECLIPTIC

EQUATOR

SUN ←

← SUN

morning   around midnight   evening

The path of the Sun and the planets along the ecliptic in July.

Calendar for July

01	18:26	Mars (mag. 1.0) 4.1°S of the Moon
02	10:07	Uranus (mag. 5.8) 4.0°S of the Moon
03–Aug.15		α-Capricornid meteor shower
03	08:28	Jupiter (mag. −2.0) 5.0°S of the Moon
03	12:01	Aldebaran 9.9°S of the Moon
05	05:06	Earth at aphelion (152,099,894 km = 1.016725 AU)
05	22:57	New Moon
06	00:04	Dwarf planet (1) Ceres at opposition (mag. 7.3)
06	15:04	Venus (mag. −3.9) 3.9°S of the Moon
06	16:29	Pollux 1.8°N of the Moon
07	18:32	Mercury (mag. −0.3) 3.2°S of the Moon
09	12:00	Regulus 3.0°S of the Moon
12–Aug.23		Southern δ-Aquariid meteor shower
12	08:11	Moon at apogee = 404,362 km
13	22:49	First Quarter
14	02:31	Spica 0.9°S of the Moon
17	20:16	Perseid meteor shower
17		Antares 0.2°S of the Moon
17–Aug.24		
21	10:17	Full Moon
22	06:39	Mercury at greatest elongation (26.9°E, mag. 0.3)
24	05:41	Moon at perigee = 364,917 km
24	20:46	Saturn (mag. 0.9) 0.4°S of the Moon
25	14:54	Neptune (mag. 7.8) 0.6°S of the Moon
28	02:51	Last Quarter
29	17:30	Uranus (mag. 5.8) 4.2°S of the Moon
30		α-Capricornid meteor shower maximum
30		Southern δ-Aquariid meteor shower maximum
30	10:37	Mars (mag. 0.9) 5.0°S of the Moon
30	17:59	Aldebaran 10.1°S of the Moon
30	23:53	Jupiter (mag. −2.1) 5.4°S of the Moon

Early morning 4:00 (BST)

Pleiades · Mars · 1 · 2 · 3 · Jupiter · Aldebaran · Menkar · ENE · E

July 1–3 • Shortly before sunrise, the Moon passes Mars, the Pleiades and Jupiter.

Evening 22:00 (BST)

Algieba · 9 · 8 · Regulus · 7 · Mercury · W · WNW

July 7–9 • On July 7, the Moon is close to Mercury. Two days later it is near Regulus and Algieba.

Evening 22:00 (BST)

Spica · Moon · SW

July 13 • The Moon is near Spica, about sixteen degrees above the southwestern horizon.

Evening 22:00 (BST)

Sabik · Antares · Moon · S

July 17 • The Moon and Antares are close together, almost due south.

Midnight 0:00 (BST)

Moon · Saturn · ESE

July 24/25 • Around midnight the Moon is near Saturn (mag. 0.9).

Early morning 4:00 (BST)

Pleiades · 30 · Mars · Jupiter · Elnath · 31 · Aldebaran · E · 25°

July 30–31 • In the early morning the Moon passes the Pleiades, Mars, Jupiter, Aldebaran and Elnath (β Tau).

JULY **75**

AQUARIUS
SATURN
ECLIPTIC
E
Circlet
PISCES
Great Square of Pegasus
PEGASUS
ANDROMEDA
ARIES
Hamal
Alpheratz
Andromeda Galaxy
M31
NGC 752
TRIANGULUM
LACERTA
Caph
Double Cluster
Algol
PERSEUS
CEPHEUS
CASSIOPEIA
Perseids
Pleiades
The Kids
Deneb
AURIGA
CYGNUS
Capella
ZENITH
CAMELOPARDALIS
Capella
M57
Ring Nebula
Polaris
LYNX
α-Aurigids
LYRA
Vega
URSA
MINOR
N
DRACO
Keystone
URSA
MAJOR
LEO MINOR
HERCULES
M13
The Plough
OPHIUCHUS
Alphecca
CANES VENATICI
SERPENS
CAPUT
CORONA
BOREALIS
M3
COMA
BERENICES
Melotte 111
BOÖTES
LIBRA
VIRGO
Arcturus
W

August 01   00:00 BST (GMT + 1hr)
August 15   23:00 BST (GMT + 1hr)
September 01 23:00 BST (GMT + 1hr)

August — Looking North

August – Looking North

A brilliant Perseid fireball, streaking alongside the Great Rift in the Milky Way, photographed in 2012 by Jens Hackmann from near Weikersheim in Germany. Four additional, fainter Perseids are also visible in the image.

Ursa Major is now the 'right way up' in the northwest, although some of the fainter stars in the south of the constellation are difficult to see. Beyond it, *Boötes* stands almost vertically in the west, but pale orange *Arcturus* is sinking towards the horizon. Higher in the sky, both **Corona Borealis** and *Hercules* are clearly visible.

In the northeast, *Capella* is clearly seen, but most of **Auriga** still remains below the horizon. Higher in the sky, **Perseus** is gradually coming into full view and, later in the night and later in the month, the beautiful **Pleiades** cluster rises above the northeastern horizon. Between Perseus and **Polaris** lies the faint and unremarkable constellation of **Camelopardalis.**

Higher still, both **Cassiopeia** and **Cepheus** are well placed for observation, despite the fact that Cassiopeia is completely immersed in the band of the Milky Way, as is the 'base' of Cepheus. **Pegasus** and **Andromeda** are now well above the eastern horizon and, below them, the constellation of **Pisces** is climbing into view. Two of the stars in the 'Summer Triangle', **Deneb** and **Vega**, are close to the zenith high overhead

Meteors

August is the month when one of the best meteor showers of the year occurs: the **Perseids**. This is a long shower, generally beginning about July 17 and continuing until around August 24, with a maximum on August 12–13, when the rate may reach as high as 100 meteors per hour (and on rare occasions, even higher). In 2024, maximum is around First Quarter, so conditions are reasonably favourable. The Perseids are debris from Comet 109P/Swift-Tuttle (the Great Comet of 1862). Perseid meteors are fast and many of the brighter ones leave persistent trains. Some bright fireballs also occur during the shower.

August — Looking South

August 01   00:00 BST (GMT + 1hr)
August 15   23:00 BST (GMT + 1hr)
September 01   22:00 BST (GMT + 1hr)

August – Looking South

The whole stretch of the summer Milky Way stretches across the sky in the south, from **Cygnus**, high in the sky near the zenith, past **Aquila**, with bright **Altair** (α Aquilae), to part of the constellation of **Sagittarius** close to the horizon, where the pattern of stars known as the 'Teapot' may be just visible. This area contains many nebulae and both open and globular clusters. Between **Albireo** (β Cygni) and Altair lie the two small constellations of **Vulpecula** and **Sagitta**, with the latter easier to distinguish (because of its shape) from the clouds of the Milky Way. Between Sagitta and **Pegasus** to the east lie the highly distinctive five stars that form the tiny constellation of **Delphinus** (again, one of the few constellations that actually bear some resemblance to the creatures after which they are named). Below Aquila, mainly in the star clouds of the Milky Way, lies **Scutum**, most famous for the bright open cluster, **M11** or the 'Wild Duck Cluster', readily visible in binoculars. To the southeast of Aquila lie the two zodiacal constellations of **Capricornus** and **Aquarius**.

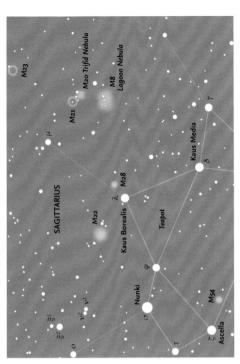

A finder chart for the gaseous nebulae M8 (the Lagoon Nebula), M20 (the Trifid Nebula) and the globular cluster M22, all in Sagittarius. Clusters M21, M23 (open) and M28 (globular) are faint. The chart shows all stars brighter than magnitude 7.5.

The Moon's phases for August 2024

New Moon

01 Day 26	02 Day 27	03 Day 28	04 11:13 Day 29	05 Day 1	06 Day 2

Full Moon

07 Day 3	08 Day 4	09 Day 5	10 Day 6	11 Day 7	12 13:19 Day 8

First Quarter

13 Day 9	14 Day 10	15 Day 11	16 Day 12		

17 Day 13	18 Day 14	19 18:26 Day 15	20 Day 16	21 Day 17	22 Day 18

Last Quarter

23 Day 19	24 Day 20	25 Day 21	26 09:26 Day 22	27 Day 23	28 Day 24

29 Day 25	30 Day 26	31 Day 27		

August – Moon and Planets

The Moon

On August 2, the Moon is 1.8°S of *Pollux*. By August 5 it is 2.9°N of *Regulus* in *Leo*. Later the same day it is 1.7°N of *Venus* (mag. -3.8). On August 6 the Moon is 7.5°N of *Mercury* and close to *Venus*. By August 10 the waxing crescent Moon is 0.7°N of *Spica*. On August 14, the Moon occults *Antares*, but this is visible only over the Pacific Ocean. By August 21, the Moon is 0.5°N of *Saturn* (mag. 0.7) and later that day is 0.7°N of *Neptune* (at mag. 7.8). By August 26 the Moon is 4.4°N of *Uranus* (mag. 5.7) and, later, 10.3°N of *Aldebaran*. On August 28, the Moon passes 5.3°N of *Mars* and on August 30, 1.7°S of *Pollux*.

The Planets

Mercury rapidly moves west and declines from mag. 0.9 to 2.2 before brightening at the end of the month to mag. 0.8. *Venus* is close by early in the month. *Mars* (mag. 0.8) is moving slowly east in *Taurus*. *Jupiter* is also moving slowly east in *Taurus*. *Saturn* (mag. 0.7) continues retrograde motion throughout August. *Uranus* (mag. 5.7) is still in *Taurus* and *Neptune* (mag. 7.8) remains in *Pisces*. Minor planet (7) Iris (mag. 8.1) is at opposition on August 6 (see the maps on page 27).

The path of the Sun and the planets along the ecliptic in August.

Calendar for August

02	23:35	Pollux 1.8°N of the Moon
04	11:13	New Moon
05	19:34	Regulus 2.9°S of the Moon
05	22:03	Venus (mag. –3.8) 1.7°S of the Moon
06	00:02	Mercury (mag. 1.6) 7.5°S of the Moon
06	15:00 *	Venus (mag. –3.8) 5.9°N of Mercury (mag. 1.6)
06	19:40	Minor planet (7) Iris at opposition (mag. 8.1)
09	01:31	Moon at apogee = 405,297 km
10	10:17	Spica 0.7°S of the Moon
12–13		Perseid meteor shower maximum
12	15:19	First Quarter
14	05:17	Antares 0.0°N of the Moon
19	18:26	Full Moon
21	03:02	Saturn (mag. 0.7) 0.5°N of the Moon
21	05:02	Moon at perigee = 360,196 km
21	22:21	Neptune (mag. 7.8) 0.7°S of the Moon
26	00:01	Uranus (mag. 5.7) 4.4°S of the Moon
26	09:26	Last Quarter
26	23:29	Aldebaran 10.3°S of the Moon
27	12:43	Jupiter (mag. –2.3) 5.7°S of the Moon
28–Sep.05		α-Aurigid meteor shower
28	00:22	Mars (mag. 0.8) 5.3°S of the Moon
30	05:25	Pollux 1.7°N of the Moon

* These objects are close together for an extended period around this time.

Early morning 4:00 (BST)

August 21 • The Moon and Saturn (mag. 0.7) are close together.

Evening 21:00 (BST)

August 13–14 • The waxing gibbous Moon passes below Antares, near the southern horizon.

Evening 21:15 (BST)

August 9–10 • Shortly after sunset, the Moon passes Spica, just seven degrees above the horizon.

Early morning 5:00 (BST)

August 30 • About thirty degrees above the eastern horizon, the crescent Moon lines up with Pollux and Castor.

Early morning 4:00 (BST)

August 26–28 • At Last Quarter, on August 26, the Moon is near the Pleiades. On August 27 and 28 it passes Jupiter and Mars. Capella and Menkalinan (α and β Aur resp.) are higher in the east.

September – Looking North

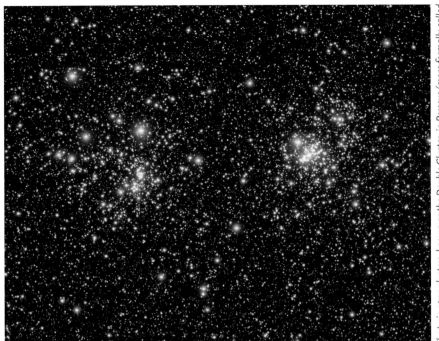

The twin open clusters, known as the Double Cluster, in Perseus (more formally called h and χ Persei) are close to the main portion of the Milky Way.

September – Looking North

The (northern) autumnal equinox occurs on September 22, when the Sun moves south of the equator in Virgo.

Ursa Major is now low in the north and to the northwest **Arcturus** and much of **Boötes** sink below the horizon later in the night and later in the month. In the northeast, **Auriga** is beginning to climb higher in the sky. Later in the month, **Taurus**, with orange **Aldebaran** (α Tauri), and even **Gemini**, with **Castor** and **Pollux**, become visible in the east and northeast. Due east, **Andromeda** is now clearly visible, with the small constellations of **Triangulum** and **Aries** (the latter a zodiacal constellation) directly below it. Practically the whole of the northern Milky Way is visible, arching across the sky, both in the north and in the south. It is not particularly clear in Auriga, or even **Perseus**, but in **Cassiopeia** and on towards **Cygnus** the clouds of stars become easier to see. The **Double Cluster** in Perseus is well placed for observation. **Cepheus** is 'upside-down' near the zenith, and the head of **Draco** and **Hercules** (with the globular cluster **M13**) beyond it are well placed for observation.

Meteors

After the major Perseid shower in August, there is very little shower activity in September. One minor, but very extended, shower, known as the **α-Aurigids**, tends to have two peaks of activity. The principal peak occurs on September 1. In 2024, the Moon is two days before New Moon, so conditions are very favourable. At maximum, however, the hourly rate hardly reaches 10 meteors per hour, although the meteors are bright and relatively easy to photograph. Activity from this shower may even extend into October. The **Southern Taurid** shower begins this month (on September 10) and, although rates are low, often produces very bright fireballs. This is a very long shower, lasting until about November 20. As a slight compensation for the lack of shower activity, however, in September the number of sporadic meteors reaches its highest rate at any time during the year.

September – Looking South

The 'Summer Triangle' is now high in the southwest, with the Great Square of **Pegasus** high in the southeast. Below Pegasus are the two zodiacal constellations of **Capricornus** and **Aquarius**. In what is otherwise an unremarkable constellation, **Algedi** (α Capricorni) is actually a visual binary, with the two stars (α¹ Cap and α² Cap) readily seen with the naked eye. **Dabih** (β Capricorni), just to the south, is also a double star, and the components are relatively easy to separate with binoculars. In Aquarius, just to the east of **Sadalmelik** (α Aquarii) there is a small asterism consisting of four stars, resembling a tiny letter 'Y', known as the 'Water Jar'. Below Aquarius is a sparsely populated area of the sky with just one bright star in the constellation of **Piscis Austrinus**. In classical illustrations, water is shown flowing from the 'Water Jar' towards bright **Fomalhaut** (α Piscis Austrini).

Another zodiacal constellation, **Pisces**, is now clearly visible to the east of Aquarius. Although faint, there is a distinctive asterism of stars, known as the 'Circlet', south of the Great Square and another line of faint stars to the east of Pegasus. Still farther down towards the horizon is the constellation of **Cetus**, with the famous variable star **Mira** (ο Ceti) at its

The constellations of Pisces and Aries. The 'Circlet' of Pisces is bottom right and the three main stars of Aries, top left. In 2024, Jupiter is in Aries until the end of April.

centre. When Mira is at maximum brightness (around mag. 3.5) it is clearly visible to the naked eye, but it disappears as it fades towards minimum (about mag. 9.5 or less). There is a finder chart for Mira on page 97.

The Moon's phases for September 2024

September – Moon and Planets

The Moon

On September 5, the waxing crescent Moon (two days after New) is 1.2°S of *Venus* (mag.-3.8). By September 17, one day before Full, the Moon is 0.3°N of *Saturn*. At Full Moon it is 0.7°N of *Neptune* (mag.7.8) On September 22, it is 4.5°N of *Uranus* (mag. 5.6) and the next day, September 23 just before Last Quarter, it is 5.7°N of brilliant *Jupiter* (mag.-2.4). On September 25 it is 4.9°N of *Mars* (mag.0.6).

The Planets

Mercury (mag. 0.9 to -1.7) reaches greatest western elongation on September 5 (see diagram on page 22). *Venus* (mag. -3.8) moves from *Virgo* into *Libra*, well east of the Sun in the evening sky. *Mars* (mag. 0.6) moves from *Taurus* into *Gemini*. *Jupiter* is moving slowly east in *Taurus*. *Saturn* is still retrograding in *Aquarius* and comes to opposition on September 8. *Uranus* (mag.5.6) begins retrograde motion on September 14 in *Aquarius*. *Neptune* (mag.7.8) is in *Pisces* and is at opposition on September 21 (see the maps on page 25).

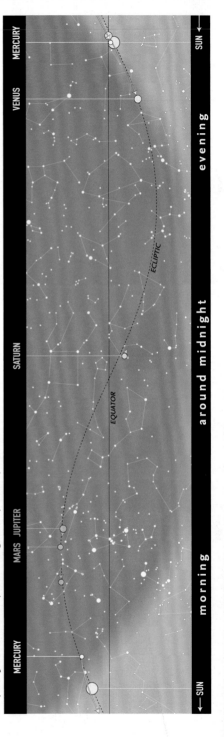

The path of the Sun and the planets along the ecliptic in September.

Calendar for September

01		α-Aurigid meteor shower maximum
01	09:16	Mercury (mag. 0.5) 5.0°S of the Moon
03	01:56	New Moon
05	02:30	Mercury at greatest elongation (18.1°W, mag. -0.3)
05	10:16	Venus (mag. -3.8) 1.2°N of the Moon
05	14:54	Moon at apogee = 406,211 km
08	04:35	Saturn at opposition (mag. 0.6)
10–Nov.20		Southern Taurid meteor shower
11	06:06	First Quarter
17	10:22	Saturn (mag. 0.6) 0.3°S of the Moon
18	02:34	Full Moon
18	02:45	Partial lunar eclipse
18	07:35	Neptune (mag. 7.8) 0.7°S of the Moon
18	13:22	Moon at perigee = 357,286 km
21	00:17	Neptune at opposition (mag. 7.8)
22	07:14	Uranus (mag. 5.6) 4.5°S of the Moon
22	12:44	September equinox
23	23:21	Jupiter (mag. -2.4) 5.7°S of the Moon
24	18:50	Last Quarter
25	11:49	Mars (mag. 0.5) 4.9°S of the Moon

Evening 20:00 (BST)

Sabik
Antares
Moon
SSW
S
SW
10°

After midnight 2:00 (BST)

Pleiades
22
23
Elnath
24
Jupiter
Aldebaran
E
ESE
—35°
10°

September 10 • *After sunset, when the Sun is only six degrees below the horizon, the Moon and Antares are very low in the south-southwest.*

September 22–24 • *High in the east, the Moon passes the Pleiades, Aldebaran, Jupiter and Elnath.*

Evening 22:00 (BST)

Saturn
16
17
Fomalhaut
Diphda
SE
SSE
10°

Early morning 3:00 (BST)

25
Mars
Alhena
Castor
26
Pollux
E
35°
10°

September 16–17 • *The almost Full Moon passes below Saturn. Fomalhaut and Diphda are only four degrees above the horizon and may not be easy to detect.*

September 25–26 • *The Moon passes Mars and Alhena. One day later it approaches Pollux & Castor.*

Morning 6:00 (BST)

Algieba
Moon
Regulus
E
10°

September 29 • *When the Moon rises in the east, it is close to Regulus and Algieba.*

October 01   00:00 BST (GMT + 1hr)
October 15   23:00 BST (GMT + 1hr)
November 01   21:00 GMT

October, Looking North

October – Looking North

Ursa Major is grazing the horizon in the north, while high overhead are the constellations of **Cepheus**, **Cassiopeia** and **Perseus**, with the Milky Way between Cepheus and Cassiopeia near the zenith. **Auriga** is now clearly visible in the east, as is **Taurus** with the **Pleiades**, **Hyades** and orange **Aldebaran**. Also in the east, **Orion** and **Gemini** are starting to rise clear of the horizon.

The constellations of **Boötes** and **Corona Borealis** are now essentially lost to view in the northwest, and **Hercules** is also descending towards the western horizon. The three stars of the Summer Triangle are still clearly visible, although **Aquila** and **Altair** are beginning to approach the horizon in the west. Towards the end of the month (October 27) Summer Time ends in Europe, with Britain reverting to Greenwich Mean Time and Europe to Central European Time.

Meteors

The **Orionids** are the major, fairly reliable meteor shower active in October. Like the May η-Aquariid shower, the Orionids are associated with Comet 1P/Halley. During this second pass through the stream of particles from the comet, slightly fewer meteors are seen than in May, but conditions are more favourable for northern observers. In both showers the meteors are very fast, and many leave persistent trains. Although the Orionid maximum is quoted as October 21–22, in fact there is a very broad maximum, lasting about a week roughly centred on that date, with hourly rates around 25. Occasionally, rates are higher (50–70 per hour). In 2024, the Moon is waning gibbous, so conditions are not particularly favourable.

The faint shower of the **Southern Taurids** (often with bright fireballs) peaks on October 10–11. The Southern Taurid maximum occurs when the Moon is around First Quarter, so conditions are more favourable

The constellation of Perseus is not only the location of the radiant of the Perseid meteor shower in August, but is also well known for the pair of open clusters, called the Double Cluster (near the top edge of the image), close to the border with Cassiopeia, and also for Algol (β Persei), the famous variable star (just below the centre of the photograph).

than those for the Orionids. Towards the end of the month (around October 20), another shower (the **Northern Taurids**) begins to show activity, which peaks early in November. The parent comet for both Taurid showers is Comet 2P/Encke. The meteors in both Taurid streams are relatively slow and bright.

A minor shower, the **Draconids**, begins on October 6 and peaks on October 8–9.

October 01   00:00 BST (GMT + 1hr)
October 15   23:00 BST (GMT + 1hr)

October – Looking South

The Great Square of *Pegasus* dominates the southern sky, framed by the two chains of stars that form the constellation of *Pisces*, together with *Alrescha* (α Piscium) at the point where the two lines of stars join. Also clearly visible is the constellation of *Cetus*, below Pegasus and Pisces. Although *Capricornus* is beginning to disappear, *Aquarius* to its east is well placed in the south, with solitary *Fomalhaut* and the constellation of *Piscis Austrinus* beneath it, close to the horizon.

The main band of the Milky Way and the Great Rift runs down from *Cygnus*, through *Vulpecula*, *Sagitta* and *Aquila* towards the western horizon. *Delphinus* and the tiny, unremarkable constellation of *Equuleus* lie between the band of the Milky Way and Pegasus. *Andromeda* is clearly visible high in the sky to the southeast, with the small constellation of *Triangulum* and the zodiacal constellation of *Aries* below it. *Perseus* is high in the east, and by now the *Pleiades* and *Taurus* are well clear of the horizon. Later in the night, and later in the month, *Orion* rises in the east, a sign that the autumn season has arrived and of the steady approach of winter.

The Moon's phases for October 2024

The constellation of Aquarius is one of the constellations that is visible in late summer and early autumn. The four stars forming the 'Y'-shape of the 'Water Jar' may be seen to the east of Sadalmelik (α Aquarii), the brightest star (top centre).

October – Moon and Planets

The Moon

On October 2 there is an annular solar eclipse, visible over the Pacific Ocean (see pages 20 and 21). The next day the Moon is 1.8°S of *Mercury* (mag.-1.6). On October 5 the Moon is 3.0°S of *Venus* (mag.-3.9). By October 14 it is 0.1°N of *Saturn* (mag.0.7). On October 15 the Moon is 0.6°N of *Neptune* (mag.7.8) and on October 19 it is 4.5°N of *Uranus* (mag.5.6). On October 23 the Moon is 3.9°N of *Mars* (mag.0.2).

The planets

Mercury (mag.-1.6) moves away from the Sun. *Venus* (mag.-3.9 to -4.0) is moving eastwards in *Virgo*. *Mars* (changing rapidly from mag.0.5 to mag.0.1) is in *Gemini*, moving towards *Cancer*. *Jupiter* (mag.-2.5 to -2.7) begins retrograde motion on October 9 and is in *Taurus*. *Saturn* (mag.0.7 to 0.8) is still retrograding slowly in *Aquarius*. *Uranus* (mag.5.6) is also retrograding in *Taurus*. *Neptune* (mag.7.8) remains in *Pisces*.

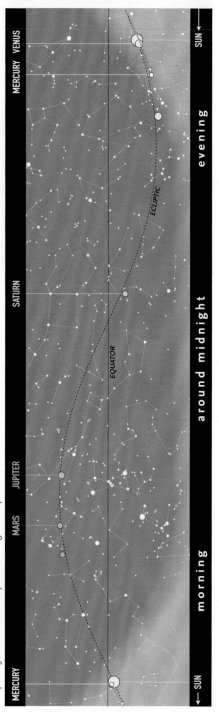

The path of the Sun and the planets along the ecliptic in October.

Calendar for October

02	18:46	Annular solar eclipse (Pacific)
02	18:49	New Moon
02	19:39	Moon at apogee = 406,516 km (most distant of the year)
02–Nov.07		Orionid meteor shower
03	00:02	Mercury (mag. –1.6) 1.8°N of the Moon
05	20:26	Venus (mag. –3.9) 3.0°N of the Moon
06–10		Draconid meteor shower
08–09		Draconid meteor shower maximum
10–11		Southern Taurid meteor shower maximum
10	18:55	First Quarter
14	18:13	Saturn (mag. 0.7) 0.1°S of the Moon
15	17:32	Neptune (mag. 7.8) 0.6°S of the Moon
17	11:26	Full Moon
17	22:50	Moon at perigee = 357,175 km
19	15:52	Uranus (mag. 5.6) 4.5°S of the Moon
20–Dec.10		Northern Taurid meteor shower
21–22		Orionid meteor shower maximum
21	08:04	Jupiter (mag. –2.6) 5.8°S of the Moon
23	19:55	Mars (mag. 0.2) 3.9°S of the Moon
24	08:03	Last Quarter
27		Summer Time ends (UK reverts to GMT)
29	22:50	Moon at apogee = 406,161 km

Evening 20:00 (BST)

October 14 • The Moon and Saturn are close together, in the southeastern sky. Fomalhaut (mag. 1.2) is only three degrees above the horizon.

After midnight 2:00 (BST)

October 23–24 • The Last Quarter Moon passes between Pollux and Mars, high in the east. Castor, Alhena and Procyon are nearby.

After midnight 2:00 (BST)

October 21 • High in the southeast, the Moon is in the company of Jupiter and Elnath, surrounded by the Pleiades, Aldebaran, Menkalinan (β Aur) and Betelgeuse.

Morning 6:00 (BST)

October 26–27 • The Moon passes between Regulus and Algieba. Denebola (β Leo) is lower and farther east.

November – Looking North

November 01 23:00 GMT
November 15 22:00 GMT
December 01 21:00 GMT

November – Looking North

The constellation of Auriga, with brilliant Capella (mag. 0.08), which, although appearing as a single star, is actually a quadruple system, consisting of a pair of yellow giant stars, gravitationally bound to a more distant pair of red dwarfs.

Most of **Aquila** has now disappeared below the horizon, but two of the stars of the 'Summer Triangle', **Vega** in **Lyra** and **Deneb** in **Cygnus**, are still clearly visible in the west. The head of **Draco** is now low in the northwest and only a small portion of **Hercules** remains above the horizon. The southernmost stars of **Ursa Major** are now coming into view. The Milky Way arches overhead, with the denser star clouds in the west and the less heavily populated region through **Auriga** and **Monoceros** in the east. High overhead, **Cassiopeia** is near the zenith and **Cepheus** has swung round to the northwest, while Auriga is now high in the northeast. **Gemini**, with **Castor** and **Pollux**, is well clear of the eastern horizon, and even **Procyon** (α Canis Minoris) is just climbing into view almost due east.

Meteors

The **Northern Taurid** shower, which began in mid-October, reaches maximum – although with just a low rate of about five meteors per hour – on November 12–13. The Moon is waxing gibbous, so conditions are not very favourable. The shower gradually trails off, ending around December 10. There is an apparent 7-year periodicity in fireball activity, but 2024 is unlikely to be a peak year. Far more striking, however, are the **Leonids**, which have a relatively short period of activity (November 6–30), with maximum on November 18. This shower is associated with Comet 55P/Tempel-Tuttle and has shown extraordinary activity on various occasions with many thousands of meteors per hour. High rates were seen in 1999, 2001 and 2002 (reaching about 3000 meteors per hour) but have fallen dramatically since then. The rate in 2024 is likely to be about 10 per hour. These meteors are the fastest shower meteors recorded (about 70 km per second) and often leave persistent trains. The shower is very rich in faint meteors. In 2024, maximum is when the Moon is waning gibbous, so conditions are not particularly favourable.

November   Looking South

S

E

N

LEO
CANCER
LYNX
M44
Praesepe
MARS
Pollux
Castor
GEMINI
Alhena
Capella
The Kids
AURIGA
Elnath
M35
CANIS MINOR
Procyon
MONOCEROS
Betelgeuse
ORION
M42
Orion Nebula
Rigel
LEPUS
CAMELOPARDALIS
Double Cluster
PERSEUS
Algol
TRIANGULUM
ZENITH
NGC 752
Hamal
ARIES
TAURUS
JUPITER
Aldebaran
Hyades
Pleiades
Northern Taurids
ERIDANUS
CEPHEUS
Caph
CASSIOPEIA
Andromeda Galaxy
M31
ANDROMEDA
Alpheratz
Great Square of Pegasus
PISCES
Alrescha
CETUS
Mira
FORNAX
SCULPTOR
HERCULES
Vega
LYRA
M57
Ring Nebula
CYGNUS
Deneb
VULPECULA
M27
Dumbbell Nebula
SAGITTA
LACERTA
PEGASUS
Circlet
ECLIPTIC
EQUULEUS
DELPHINUS
AQUILA
Altair
M15
AQUARIUS
Water Jar
SATURN
CAPRICORNUS
PISCIS AUSTRINUS
Fomalhaut

November – Looking South

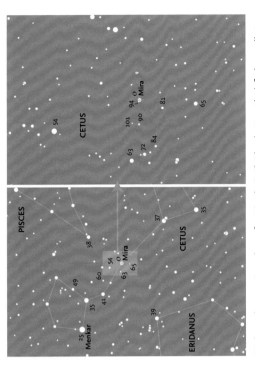

Orion has now risen above the eastern horizon, and part of the long, straggling constellation of **Eridanus** (which begins near **Rigel**) is visible to the west of Orion. Higher in the sky, **Taurus**, with the **Pleiades** cluster, and orange **Aldebaran** are now easy to observe. To their west, both **Pisces** and **Cetus** are close to the meridian. The famous long-period variable star, **Mira** (o Ceti), with a typical range of magnitude 3.4 to 9.8, is favourably placed for observation. In the southwest, **Capricornus** has slipped below the horizon, but **Aquarius** remains visible. Even farther west, **Altair** may be seen early in the night, but most of **Aquila** has already disappeared from view. **Delphinus**, together with **Sagitta** and **Vulpecula** in the Milky Way, will soon vanish for another year. Both **Pegasus** and **Andromeda** are easy to see, and one of the lines of stars that make up Andromeda finishes close to the zenith, which is also close to one of the outlying stars of **Perseus**, high in the east.

Finder and comparison charts for Mira (o Ceti). The chart on the left shows all stars brighter than magnitude 6.5. The chart on the right shows stars down to magnitude 10.0. The comparison star magnitudes are shown without the decimal point.

The Moon's phases for November 2024

November – Moon and Planets

The Moon

On November 3, the Moon is 2.1°S of **Mercury** (mag. -0.3). On November 4 the Moon is 0.1°S of **Antares** and the next day (at 00:15) is 3.1°S of **Venus** (mag. -4.0). By November 12 the Moon is 0.6°N of **Neptune** (mag. 7.8). On November 16 the Moon is 4.4°N of **Uranus** (mag. 5.6). On November 17 the Moon is 10.3°N of **Aldebaran**, later that day it is 5.5°N of **Jupiter** (mag. -2.8). By November 20, the Moon is 1.9°S of **Pollux**, later that day it is 2.4°N of **Mars** in **Cancer**. By November 22, the Moon passes 2.7°N of **Regulus** between it and **Algieba**. On November 27 the Moon is 0.4°N of **Spica** in **Virgo**.

The Planets

Mercury reaches eastern elongation on November 16. **Venus** is very bright (mag. -4.0 to -4.2) and moving eastwards away from the Sun. **Mars** (mag. 0.1 to -0.7) moves eastwards in **Cancer**. **Jupiter** (mag. -2.7 to -2.8) is retrograding in **Taurus**. **Saturn** (mag. 0.8 to 1.0) is in **Aquarius**. **Uranus** (mag. 5.6) is in **Taurus** and comes to opposition on November 17 (see the maps on page 25). **Neptune** (mag. 7.8) remains in **Pisces**.

The path of the Sun and the planets along the ecliptic in November.

Calendar for November

01	12:47	New Moon
03	07:36	Mercury (mag.–0.3) 2.1°N of the Moon
04	01:06	Antares 0.1°N of the Moon
05	00:15	Venus (mag.–4.0) 3.1°N of the Moon
06–30		Leonid meteor shower
09	05:55	First Quarter
11	01:43	Saturn (mag. 0.9) 0.1°S of the Moon
12–13		Northern Taurid meteor shower Maximum
12	02:25	Neptune (mag.7.8) 0.6°S of the Moon
14	11:16	Moon at perigee = 360,109 km
15	21:28	Full Moon
16	01:13	Uranus (mag. 5.6) 4.4°S of the Moon
16	08:09	Mercury at greatest elongation (22.6°E, mag.–0.3)
17	02:02	Aldebaran 10.3°S of the Moon
17	02:45	Uranus at opposition (mag.5.6)
17	14:53	Jupiter (mag.–2.8) 5.5°S of the Moon
18		Leonid meteor shower maximum
20	02:44	Pollux 1.9°N of the Moon
20	21:09	Mars (mag.–0.3) 2.4°S of the Moon
22	21:29	Regulus 2.7°S of the Moon
23	01:28	Last Quarter
26	11:56	Moon at apogee = 405,314 km
27	12:16	Spica 0.4°S of the Moon

Evening 17:00

November 5 • The Moon and Venus are low in the south-southwest, flanked by Sabik and Nunki.

Evening 17:00

November 10–11 • The waxing gibbous Moon passes Saturn in the southeast.

Early morning 4:00

November 15–16 • The Full Moon passes Hamal, Uranus (mag. 5.6) and the Pleiades. Aldebaran is nearby.

Evening 20:00

November 16–17 • The Moon passes Aldebaran, Jupiter and Elnath, almost due east.

Evening 23:00

November 19–20 • High in the eastern sky the Moon moves from Pollux (mag. 1.2) to Mars (mag. –0.3).

Morning 6:00

November 27 • In the southeastern sky the Moon and Spica are close together.

December – Looking North

December 01   23:00 GMT
December 15   22:00 GMT
January 01   21:00 GMT

December – Looking North

Ursa Major has now swung around and is starting to 'climb' in the east. The fainter stars in the southern part of the constellation are now fully in view. The other bear, *Ursa Minor*, 'hangs' below *Polaris* in the north. Directly above it is the faint constellation of *Camelopardalis*, with the other inconspicuous circumpolar constellation, *Lynx*, to its east. *Vega* (α Lyrae) is skimming the horizon in the northwest, but *Deneb* (α Cygni) and most of *Cygnus* remain visible farther west. In the east, *Regulus* (α Leonis) and the constellation of *Leo* are beginning to rise above the horizon. *Cancer* stands high in the east, with *Gemini* even higher in the sky. *Perseus* is at the zenith, with *Auriga* and *Capella* between it and Gemini. Because it is so high in the sky, now is a good time to examine the star clouds of the fainter portion of the Milky Way, between *Cassiopeia* in the west to Gemini and *Orion* in the east.

The constellation of Andromeda largely consists of a line of bright stars running northeast from α Andromedae, Alpheratz (bottom right), which is one of the stars forming the Great Square of Pegasus. The small constellation of Triangulum appears on the left-hand (eastern) side, below Andromeda.

with a rate at maximum of 5–10, occasionally rising to 25 per hour. Maximum in 2024 occurs when the Moon is around Last Quarter, so conditions are moderately favourable. The parent body is Comet 8P/Tuttle.

Meteors

There is one significant meteor shower in December (the last major shower of the year). This is the *Geminid* shower, which is visible over the period December 4–20 and comes to maximum on December 14–15, when the Moon is Full, so conditions are particularly unfavourable. It is one of the most active showers of the year, and in some years is the strongest, with a peak rate of around 100 meteors per hour. It is the one major shower that shows good activity before midnight. The meteors have been found to have a much higher density than other meteors (which are derived from cometary material). It was eventually established that the Geminids and the asteroid Phaethon had similar orbits. So the Geminids are assumed to consist of denser, rocky material. They are slower than most other meteors and often appear to last longer. The brightest often break up into numerous luminous fragments that follow similar paths across the sky. There is a second shower: the *Ursids*, active December 17–26, peaking on December 23,

CYGNUS

EQUULEUS

M15

Water Jar

AQUARIUS

PEGASUS

SATURN

Circlet

Great Square of Pegasus

ECLIPTIC

CEPHEUS

LACERTA

Alpheratz

ANDROMEDA

PISCES

CETUS

SCULPTOR

CASSIOPEIA

Caph

Andromeda Galaxy M31

NGC 752

Hamal

Mira

FORNAX

Double Cluster

PERSEUS

TRIANGULUM

ARIES

Algol

Pleiades

ZENITH

Hyades

TAURUS

ERIDANUS

S

CAMELOPARDALIS

The Kids

Capella

AURIGA

Elnath

JUPITER

Aldebaran

ORION

LEPUS

COLUMBA

LYNX

M35

Betelgeuse

Rigel

M42
Orion Nebula

Geminids

Castor

MONOCEROS

Pollux

Alhena

CANIS MINOR

Sirius

CANIS MAJOR

URSA MAJOR

GEMINI

Procyon

CANCER

MARS

LEO MINOR

M44
Praesepe

M67

Algieba

LEO

HYDRA

Regulus

E

December – Looking South

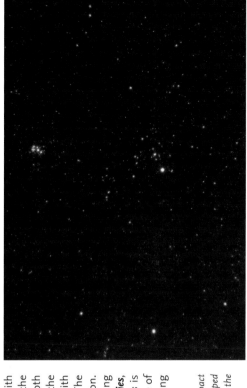

The fine open cluster of the **Pleiades** is due south around 22:00, with the **Hyades** cluster, **Aldebaran** and the rest of **Taurus** clearly visible to the east. **Auriga** (with **Capella**) and **Gemini** (with **Castor** and **Pollux**) are both well-placed for observation. **Orion** has made a welcome return to the winter sky, and both **Canis Minor** (with **Procyon**) and **Canis Major** (with **Sirius**, the brightest star in the sky) are now well above the horizon. The small, poorly known constellation of **Lepus** lies to the south of Orion. In the west, **Aquarius** has now disappeared, and **Cetus** is becoming lower, but **Pisces** is still easily seen, as are the constellations of **Aries**, **Triangulum** and **Andromeda** above it. The Great Square of **Pegasus** is starting to plunge down towards the western horizon, and because of its orientation on the sky, appears more like a large diamond, standing on one point, than a square.

The constellation of Taurus contains two contrasting open clusters: the compact Pleiades, with its striking blue-white stars, and the more scattered, 'V'-shaped Hyades, which are much closer to us. Orange Aldebaran (α Tauri) is not related to the Hyades, but lies between it and the Earth.

The Moon's phases for December 2024

December – Moon and Planets

The Moon

On December 1, the Moon is 0.1°S of **Antares** in **Scorpius**. The next day it is 5.0°S of **Mercury** (too close to the Sun to be visible). By December 4 the Moon is 2.3°S of **Venus** (mag. -4.2) in **Sagittarius**. On December 8 it is 0.3°N of **Saturn** in **Aquarius**. The Moon passes 0.8°N of **Neptune** on December 9 and 4.4°N of **Uranus** on December 13. The next day, December 14, the Moon is 10.3°N of **Aldebaran** and later, 5.5°N of **Jupiter** (mag. -2.8). By December 17 the Moon is 2.1°S of **Pollux**. On December 18 the Moon is 0.9°N of **Mars**, which has begun to retrograde in **Cancer**. On December 20, the Moon passes 2.4°N of **Regulus**. On December 24 it is 0.2°N of **Spica** and on December 28 0.1°S of **Antares**. One day before New, the Moon is 6.4°S of **Mercury**.

The planets

Mercury reaches western elongation on December 25 (see the diagram on page 22). **Venus** (mag. -4.2 to -4.5) moves rapidly across Capricornus. **Mars** begins retrograde motion on December 9. **Jupiter** (mag. -2.8) comes to opposition on December 7. **Saturn** (mag. 1.0 to 1.1) is in **Aquarius**. **Uranus** (mag. 5.6) is in **Taurus**, and **Neptune** (mag. 7.9) remains in **Pisces**, where it has been all year.

The path of the Sun and the planets along the ecliptic in December.

Calendar for December

01	06:21	New Moon
01	07:26	Antares 0.1°N of the Moon
02	02:09	Mercury (mag. 2.7) 5.0°N of the Moon
04–20		Geminid meteor shower
04	22:40	Venus (mag. -4.2) 2.3°N of the Moon
07	20:58	Jupiter at opposition (mag. -2.8)
08	09:19	Saturn (mag. 1.0) 0.3°S of the Moon
8	15:27	First Quarter
09	09:19	Neptune (mag. 7.9) 0.8°S of the Moon
2	13:20	Moon at perigee = 365,361 km
3	09:34	Uranus (mag. 5.6) 4.4°S of the Moon
4–15		Geminid meteor shower maximum
4	05:31	Minor planet (15) Eunomia at opposition (mag. 8.0)
4	12:30	Aldebaran 10.3°S of the Moon
4	19:32	Jupiter (mag. -2.8) 5.5°S of the Moon
5	09:02	Full Moon
7–26		Ursid meteor shower
7	12:49	Pollux 2.1°N of the Moon
8	08:49	Mars (mag. -0.9) 0.9°S of the Moon
0	06:18	Regulus 2.4°S of the Moon
1	09:21	Northern winter solstice
2	22:18	Last Quarter
3		Ursid meteor shower maximum
4	07:25	Moon at apogee = 404,485 km
4	20:11	Spica 0.2°S of the Moon
5	02:30	Mercury at greatest elongation (22.0°W, mag. -0.4)
8	15:17	Antares 0.1°N of the Moon
9	04:21	Mercury (mag. -0.4) 6.4°N of the Moon
		New Moon

Evening 17:00

December 4–5 • *After sunset, the Moon passes below brilliant Venus (mag. -4.0). Nunki is only two degrees above the horizon and is probably lost in dusk.*

December 7–8 • *In the southern sky, the First Quarter Moon passes below Saturn. Fomalhaut is twenty degrees closer to the horizon.*

Evening 19:30

December 13–15 • *The Moon is nearly Full, when it passes the Pleiades, Aldebaran, Jupiter and Elnath.*

Evening 21:00

December 17 • *Mars, the Moon, Pollux and Castor form a nice curved line in the eastern sky.*

Evening 23:00

December 19–20 • *Due east, the Moon passes between Regulus and Algieba.*

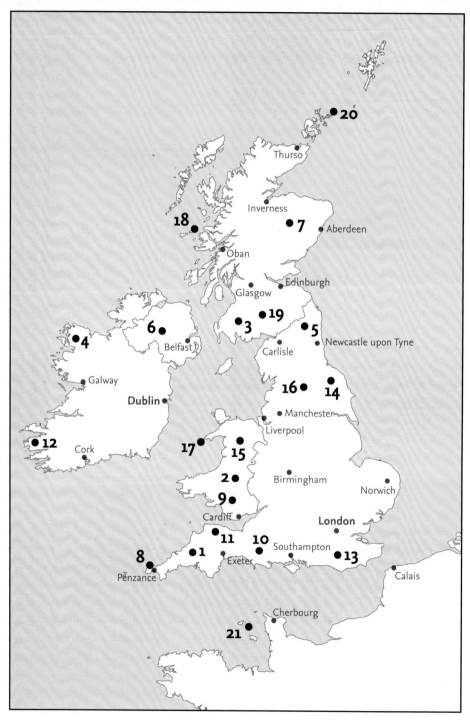

Dark Sky Sites

International Dark-Sky Association Sites

The *International Dark-Sky Association* (IDA) recognizes various categories of sites that offer areas where the sky is dark at night, free from light pollution and particularly suitable for astronomical observing. A number of sites in Great Britain and Ireland have been given specific recognition and are shown on the map. These are:

Parks

1   *Bodmin Moor Dark Sky Landscape*
2   *Elan Valley Estate*
3   *Galloway Forest Park*
4   *Mayo Dark Sky Park*
5   *Northumberland National Park and Kielder Water & Forest Park*
6   *OM Dark Sky Park & Observatory*
7   *Tomintoul and Glenlivet – Cairngorms*
8   *West Penwith*

Reserves

9   *Brecon Beacons National Park*
10   *Cranborne Chase*
11   *Exmoor National Park*
12   *Kerry*
13   *Moore's Reserve South Down National Park*
14   *North York Moors National Park*
15   *Snowdonia National Park*
16   *Yorkshire Dales National Park*

Sanctuaries

17   *Ynys Enlli (Bardsey Island)*

Communities

17   *The island of Coll (Inner Hebrides, Scotland)*
18   *Moffat*
19   *North Ronaldsay Dark Sky Island (Orkney Islands, Scotland)*
20   *The island of Sark (Channel Islands)*

Details of these sites and web links may be found at the IDA website: https://www.darksky.org/ Many of these sites have major observatories or other facilities available for public observing.

Dark Sky Discovery Sites

In Britain there is also the *Dark Sky Discovery* organisation. This gives recognition to smaller sites, again free from immediate light pollution, that are open to observing at any time. . A full listing of sites is at: https://www.darkskydiscovery.org.uk/

Glossary and Tables

aphelion	The point on an orbit that is farthest from the Sun.
apogee	The point on its orbit at which the Moon is farthest from the Earth.
appulse	The apparently close approach of two celestial objects; two planets, or a planet and star.
astronomical unit	(AU) The mean distance of the Earth from the Sun, 149,597,870 km.
celestial equator	The great circle on the celestial sphere that is in the same plane as the Earth's equator.
celestial sphere	The apparent sphere surrounding the Earth on which all celestial bodies (stars, planets, etc.) seem to be located.
conjunction	The point in time when two celestial objects have the same celestial longitude. In the case of the Sun and a planet, superior conjunction occurs when the planet lies on the far side of the Sun (as seen from Earth). For Mercury and Venus, inferior conjunction occurs when they pass between the Sun and the Earth.
direct motion	Motion from west to east on the sky.
ecliptic	The apparent path of the Sun across the sky throughout the year. Also: the plane of the Earth's orbit in space.
elongation	The point at which an inferior planet has the greatest angular distance from the Sun, as seen from Earth.
equinox	The two points during the year when night and day have equal duration. Also: the points on the sky at which the ecliptic intersects the celestial equator. The vernal (spring) equinox is of particular importance in astronomy.
gibbous	The stage in the sequence of phases at which the illumination of a body lies between half and full. In the case of the Moon, the term is applied to phases between First Quarter and Full, and between Full and Last Quarter.
inferior planet	Either of the planets Mercury or Venus, which have orbits inside that of the Earth.
magnitude	The brightness of a star, planet or other celestial body. It is a logarithmic scale, where larger numbers indicate fainter brightness. A difference of 5 in magnitude indicates a difference of 100 in actual brightness, thus a first-magnitude star is 100 times as bright as one of sixth magnitude.
meridian	The great circle passing through the North and South Poles of a body and the observer's position; or the corresponding great circle on the celestial sphere that passes through the North and South Celestial Poles and also through the observer's zenith.
nadir	The point on the celestial sphere directly beneath the observer's feet, opposite the zenith.
occultation	The disappearance of one celestial body behind another, such as when stars or planets are hidden behind the Moon.
opposition	The point on a superior planet's orbit at which it is directly opposite the Sun in the sky.
perigee	The point on its orbit at which the Moon is closest to the Earth.
perihelion	The point on an orbit that is closest to the Sun.
retrograde motion	Motion from east to west on the sky.
superior planet	A planet that has an orbit outside that of the Earth.
vernal equinox	The point at which the Sun, in its apparent motion along the ecliptic, crosses the celestial equator from south to north. Also known as the First Point of Aries.
zenith	The point directly above the observer's head.
zodiac	A band, streching 8° on either side of the ecliptic, within which the Moon and planets appear to move. It consists of twelve equal areas, originally named after the constellation that once lay within it.

The Greek Alphabet

α	Alpha	ε	Epsilon	ι	Iota	ν	Nu	ρ	Rho	φ (φ)	Phi
β	Beta	ζ	Zeta	κ	Kappa	ξ	Xi	σ (ς)	Sigma	χ	Chi
γ	Gamma	η	Eta	λ	Lambda	ο	Omicron	τ	Tau	ψ	Psi
δ	Delta	θ (ϑ)	Theta	μ	Mu	π	Pi	υ	Upsilon	ω	Omega

The Constellations

There are 88 constellations covering the whole of the celestial sphere, but 24 of these in the southern hemisphere can never be seen (even in part) from the latitude of Britain and Ireland, so are omitted from this table. The names themselves are expressed in Latin, and the names of stars are frequently given by Greek letters followed by the genitive of the constellation name. The genitives and English names of the various constellations are included.

Name	Genitive	Abbr.	English name
Andromeda	Andromedae	And	Andromeda
Antlia	Antliae	Ant	Air Pump
Aquarius	Aquarii	Aqr	Water Bearer
Aquila	Aquilae	Aql	Eagle
Aries	Arietis	Ari	Ram
Auriga	Aurigae	Aur	Charioteer
Boötes	Boötis	Boo	Herdsman
Camelopardalis	Camelopardalis	Cam	Giraffe
Cancer	Cancri	Cnc	Crab
Canes Venatici	Canum Venaticorum	CVn	Hunting Dogs
Canis Major	Canis Majoris	CMa	Big Dog
Canis Minor	Canis Minoris	CMi	Little Dog
Capricornus	Capricorni	Cap	Sea Goat
Cassiopeia	Cassiopeiae	Cas	Cassiopeia
Centaurus	Centauri	Cen	Centaur
Cepheus	Cephei	Cep	Cepheus
Cetus	Ceti	Cet	Whale
Columba	Columbae	Col	Dove
Coma Berenices	Comae Berenices	Com	Berenice's Hair
Corona Australis	Coronae Australis	CrA	Southern Crown
Corona Borealis	Coronae Borealis	CrB	Northern Crown
Corvus	Corvi	Crv	Crow
Crater	Crateris	Crt	Cup
Cygnus	Cygni	Cyg	Swan
Delphinus	Delphini	Del	Dolphin
Draco	Draconis	Dra	Dragon
Equuleus	Equulei	Equ	Little Horse
Eridanus	Eridani	Eri	River Eridanus
Fornax	Fornacis	For	Furnace
Gemini	Geminorum	Gem	Twins
Hercules	Herculis	Her	Hercules
Hydra	Hydrae	Hya	Water Snake

Name	Genitive	Abbr.	English name
Lacerta	Lacertae	Lac	Lizard
Leo	Leonis	Leo	Lion
Leo Minor	Leonis Minoris	LMi	Little Lion
Lepus	Leporis	Lep	Hare
Libra	Librae	Lib	Scales
Lupus	Lupi	Lup	Wolf
Lynx	Lyncis	Lyn	Lynx
Lyra	Lyrae	Lyr	Lyre
Microscopium	Microscopii	Mic	Microscope
Monoceros	Monocerotis	Mon	Unicorn
Ophiuchus	Ophiuchi	Oph	Serpent Bearer
Orion	Orionis	Ori	Orion
Pegasus	Pegasi	Peg	Pegasus
Perseus	Persei	Per	Perseus
Pisces	Piscium	Psc	Fishes
Piscis Austrinus	Piscis Austrini	PsA	Southern Fish
Puppis	Puppis	Pup	Stern
Pyxis	Pyxidis	Pyx	Compass
Sagitta	Sagittae	Sge	Arrow
Sagittarius	Sagittarii	Sgr	Archer
Scorpius	Scorpii	Sco	Scorpion
Sculptor	Sculptoris	Scl	Sculptor
Scutum	Scuti	Sct	Shield
Serpens	Serpentis	Ser	Serpent
Sextans	Sextantis	Sex	Sextant
Taurus	Tauri	Tau	Bull
Triangulum	Trianguli	Tri	Triangle
Ursa Major	Ursae Majoris	UMa	Great Bear
Ursa Minor	Ursae Minoris	UMi	Lesser Bear
Vela	Velorum	Vel	Sails
Virgo	Virginis	Vir	Virgin
Vulpecula	Vulpeculae	Vul	Fox

Some common asterisms

Belt of Orion	δ, ε and ζ Orionis
Big Dipper	α, β, γ, δ, ε, ζ and η Ursae Majoris
Circlet	γ, θ, ι, λ and κ Piscium
Guards (or Guardians)	β and γ Ursae Minoris
Head of Cetus	α, γ, ξ², μ and λ Ceti
Head of Draco	β, γ, ξ and ν Draconis
Head of Hydra	δ, ε, ζ, η, ρ and σ Hydrae
Keystone	ε, ζ, η and π Herculis
Kids	ζ and η Aurigae
Little Dipper	β, γ, η, ζ, ε, δ and α Ursae Minoris
Lozenge	= Head of Draco
Milk Dipper	ζ, γ, σ, φ and λ Sagittarii
Plough or Big Dipper	α, β, γ, δ, ε, ζ and η Ursae Majoris
Pointers	α and β Ursae Majoris
Sickle	α, η, γ, ζ, μ and ε Leonis
Square of Pegasus	α, β and γ Pegasi with α Andromedae
Sword of Orion	θ and ι Orionis
Teapot	γ, ε, δ, λ, φ, σ, τ and ζ Sagittarii
Wain (or Charles' Wain)	= Plough
Water Jar	γ, η, κ and ζ Aquarii
Y of Aquarius	= Water Jar

Acknowledgements

Denis Buczynski, Portmahomack, Ross-shire – p.32 (Fireball), p.35 (Quadrantid fireball)
Stephen Edberg – pp.35, 53, 89, 91, 103 (Constellation photographs)
Akira Fuji – p.21 (Lunar eclipse)
Jens Hackmann, Bad Mergentheim, Germany – p.77 (Perseid fireball)
Bernhard Hubl – pp.49, 53, 55, 71, 85, 95 (Constellation photographs)
Nick James – p.29 (Comet NEOWISE)
Damian Peach – p.23 (Mars photo)
peresanz/Shutterstock – p.37 (Orion)
Ken Sperber, California – p.83 (Double Cluster)
Wil Tirion, Capelle aan den IJssel, The Netherlands – p.43, 67, 101 (Constellation photographs)
Alan Tough, Nairn, Highland, Scotland – p.65 (noctilucent clouds)
For the 2024 Guide, specialist editorial support was provided by Jessica Lee, Astronomy Education Officer, and Patricia Skelton, Senior Astronomy Manager: Education & Outreach at Royal Observatory Greenwich.

Further Information

Books

Bone, Neil (1993), *Observer's Handbook: Meteors*, George Philip, London & Sky Publ. Corp., Cambridge, Mass.

Cook, J., ed. (1999), *The Hatfield Photographic Lunar Atlas*, Springer-Verlag, New York

Dunlop, Storm (1999), *Wild Guide to the Night Sky*, HarperCollins, London

Dunlop, Storm (2012), *Practical Astronomy*, 3rd edn, Philip's, London

Dunlop, Storm, Rükl, Antonin & Tirion, Wil (2005), *Collins Atlas of the Night Sky*, HarperCollins, London

Heifetz, Milton & Tirion, Wil (2017), *A Walk through the Heavens*, 4th edn, Cambridge University Press, Cambridge

O'Meara, Stephen J. (2008), *Observing the Night Sky with Binoculars*, Cambridge University Press, Cambridge

Ridpath, Ian (2018), *Star Tales*, 2nd edn, Lutterworth Press, Cambridge, UK

Ridpath, Ian, ed. (2003), *Oxford Dictionary of Astronomy*, 2nd edn, Oxford University Press, Oxford

Ridpath, Ian, ed. (2004), *Norton's Star Atlas*, 20th edn, Pi Press, New York

Ridpath, Ian & Tirion, Wil (2004), *Collins Gem – Stars*, HarperCollins, London

Ridpath, Ian & Tirion, Wil (2017), *Collins Pocket Guide Stars and Planets*, 5th edn, HarperCollins, London

Ridpath, Ian & Tirion, Wil (2019), *The Monthly Sky Guide*, 10th edn, Dover Publications, New York

Rükl, Antonín (1990), *Hamlyn Atlas of the Moon*, Hamlyn, London & Astro Media Inc., Milwaukee

Rükl, Antonín (2004), *Atlas of the Moon*, Sky Publishing Corp., Cambridge, Mass.

Scagell, Robin (2000), *Philip's Stargazing with a Telescope*, George Philip, London

Stimac, Valerie (2019), *A Practical Guide to Astrotourism*, Lonely Planet, Franklin, TN

Tirion, Wil (2011), *Cambridge Star Atlas*, 4th edn, Cambridge University Press, Cambridge

Tirion, Wil & Sinnott, Roger (1999), *Sky Atlas 2000.0*, 2nd edn, Sky Publishing Corp., Cambridge, Mass. & Cambridge University Press, Cambridge

Journals

Astronomy, Astro Media Corp., 21027 Crossroads Circle, P.O. Box 1612, Waukesha, WI 53187-1612 USA. http://astronomy.com

Astronomy Now, Pole Star Publications, PO Box 175, Tonbridge, Kent TN10 4QX UK. http://astronomynow.com

Sky at Night Magazine, BBC publications, London. http://skyatnightmagazine.com

Sky & Telescope, Sky Publishing Corp., Cambridge, MA 02138-1200 USA. http://www.skyandtelescope.com/

Societies

British Astronomical Association, Burlington House, Piccadilly, London W1J 0DU. http://www.britastro.org/

The principal British organization for amateur astronomers (with some professional members), particularly for those interested in carrying out observational programmes. Its membership is, however, worldwide. It publishes fully refereed, scientific papers and other material in its well-regarded journal.

Federation of Astronomical Societies, Secretary: Ken Sheldon, Whitehaven, Maytree Road, Lower Moor, Pershore, Worcs. WR10 2NY. http://www.fedastro.org.uk/fas/

An organization that is able to provide contact information for local astronomical societies in the United Kingdom.

Royal Astronomical Society, Burlington House, Piccadilly, London W1J 0BQ. http://www.ras.org.uk/
The premier astronomical society, with membership primarily drawn from professionals and experienced amateurs. It has an exceptional library and is a designated centre for the retention of certain classes of astronomical data. Its publications are the standard medium for dissemination of astronomical research.

Society for Popular Astronomy, 36 Fairway, Keyworth, Nottingham NG12 5DU.
http://www.popastro.com/
A society for astronomical beginners of all ages, which concentrates on increasing members' understanding and enjoyment, but which does have some observational programmes. Its journal is entitled *Popular Astronomy*.

Software

Planetary, Stellar and Lunar Visibility (planetary and eclipse freeware): Alcyone Software, Germany.
http://www.alcyone.de
Redshift, Redshift-Live. http://www.redshift-live.com/en/
Starry Night & Starry Night Pro, Sienna Software Inc., Toronto, Canada. http://www.starrynight.com

Internet sources

There are numerous sites with information about all aspects of astronomy, and all of those have numerous links. Although many amateur sites are excellent, treat any statements and data with caution. The sites listed below offer accurate information. Please note that the URLs may change. If so, use a good search engine, such as Google, to locate the information source.

Information

Astronomical data (inc. eclipses) HM Nautical Almanac Office: http://astro.ukho.gov.uk
Auroral information Michigan Tech: http://www.geo.mtu.edu/weather/aurora/
Comets JPL Solar System Dynamics: http://ssd.jpl.nasa.gov/
American Meteor Society: http://amsmeteors.org/
Deep-sky objects Saguaro Astronomy Club Database: http://www.virtualcolony.com/sac/
Eclipses NASA Eclipse Page: http://eclipse.gsfc.nasa.gov/eclipse.html
Moon (inc. Atlas) Inconstant Moon: http://www.inconstantmoon.com/
Planets Planetary Fact Sheets: http://nssdc.gsfc.nasa.gov/planetary/planetfact.html
Satellites (inc. International Space Station)
Heavens Above: http://www.heavens-above.com/
Visual Satellite Observer: http://www.satobs.org/
Star Chart: http://www.skyandtelescope.com/observing/interactive-sky-watching-tools/interactive-sky-chart/
What's Visible
Skyhound: http://www.skyhound.com/sh/skyhound.html
Skyview Cafe: http://www.skyviewcafe.com

Institutes and Organizations

European Space Agency: http://www.esa.int/
International Dark-Sky Association: http://www.darksky.org/
International Meteor Organization https://www.imo.net/
RASC Dark Sky: https://rasc.ca/lpa/dark-sky-sites/
Jet Propulsion Laboratory: http://www.jpl.nasa.gov/
Lunar and Planetary Institute: http://www.lpi.usra.edu/
National Aeronautics and Space Administration: http://www.hq.nasa.gov/
Solar Data Analysis Center: http://umbra.gsfc.nasa.gov/
Space Telescope Science Institute: http://www.stsci.edu/